Interdisciplinary Mathematics — Volume 29

LIE - THEORETIC ODE NUMERICAL ANALYSIS, MECHANICS AND DIFFERENTIAL SYSTEMS

ROBERT HERMANN

MATH SCI PRESS

Copyright © 1994 by Robert Hermann
Al rights reserved

ISBN 0-915692-45-7

Math Sci Press
53 Jordan Road
Brookline, Massachusetts

Library of Congress Cataloging-in-Publication Data

Hermann, Robert, 1931-
    Lie-theoretic ODE numerical analysis, mechanics, and differential systems / Robert Hermann.
       p.   cm. -- (Interdisciplinary mathematics ; v. 29)
    Includes bibliographical references.
    ISBN 0-915692-45-7
    1. Differential equations--Numerical solutions. 2. Numerical analysis. 3. Lie groups. I. Title. II. Series: Hermann, Robert, 1931-   Interdisciplinary mathematics ; v. 29.
QA372.H485 1994
515'.35--dc20   94-10816
                                                                                      CIP

MATH SCI PRESS

53 Jordan Road, Brookline, MA 02146 (USA) (617) 738-0307

INTERDISCIPLINARY MATHEMATICS, by Robert Hermann

1. GENERAL ALGEBRAIC IDEAS. 205 pages. 1973.
2. LINEAR AND TENSOR ALGEBRA. 185 pages. 1973.
3. ALGEBRAIC TOPICS IN SYSTEM THEORY. 177 pages.
4. ENERGY MOMENTUM TENSORS. 153 pages. 1973.
5. TOPICS IN GENERAL RELATIVITY. 161 pages. 1973.
6. TOPICS IN THE MATHEMATICS OF QUANTUM MECHANICS, 250 pages. 1973.
7. SPINORS, CLIFFORD AND CAYLEY ALGEBRAS. 276 pages.
8. LINEAR SYSTEMS AND INTRODUCTORY ALGEBRAIC GEOMETRY. 282 pages. 1974.
9. GEOMETRIC STRUCTURE OF SYSTEMS-CONTROL THEORY AND PHYSICS, part A. 450 pages. 1974.
10. GAUGE FIELDS AND CARTAN-EHRESMANN CONNECTIONS. 515 pages. 1975.
11. GEOMETRIC STRUCTURE OF SYSTEMS-CONTROL THEORY AND PHYSICS, part B. 484 pages 1976.
12. GEOMETRIC THEORY OF NON-LINEAR DIFFERENTIAL EQUATIONS, BAEKLUND TRANSFORMATIONS AND SOLITONS, part A, 313 pages. 1976.
13. ALGEBRO-GEOMETRIC AND LIE-THEORETIC TECHNIQUES IN SYSTEM THEORY, with C. Martin, 1977. 256 pages.
14. GEOMETRIC THEORY OF NON-LINEAR DIFFERENTIAL EQUATIONS, BAEKLUND TRANSFORMATIONS AND SOLITONS, part B. 1976. 336 pages.
15. TODA LATTICES, COSYMPLECTIC MANIFOLDS, BAEKLUND TRANSFORMATIONS AND KINKS, part A. 1977. 225 pages. .
16. QUANTUM AND FERMION DIFFERENTIAL GEOMETRY. 1977. 196 pages.
17. DIFFERENTIAL GEOMETRY AND THE CALCULUS OF VARIATIONS, 2nd EDITION. 1977. 724 pages.
18. TODA LATTICES, COSYMPLECTIC MANIFOLDS, BAEKLUND TRANSFORMATIONS AND KINKS, part B. 1977. 145 pages. 6
19. YANG-MILLS, KALUZA-KLEIN AND THE EINSTEIN PROGRAM. 1978. 198 pages.
20. CARTANIAN GEOMETRY, NONLINEAR WAVES, AND CONTROL THEORY, part A. 1980. 501 pages.
21. CARTANIAN GEOMETRY, NONLINEAR WAVES, AND CONTROL THEORY, part B. 1980. 585 pages.
22. TOPICS IN THE GEOMETRIC THEORY OF LINEAR SYSTEMS. 1984. 281 pages.
23. TOPICS IN THE GEOMETRIC THEORY OF INTEGRABLE SYSTEMS. 1984. 347 pages.
24. TOPICS IN PHYSICAL GEOMETRY. 595 pages. 1988.
25. GEOMETRIC COMPUTING SCIENCE: FIRST STEPS. 1991.
26. GEOMETRIC STRUCTURES IN NONLINEAR PHYSICS. 1991.
27. CONSTRAINED MECHANICS AND LIE THEORY. 1992.
28. LIE -CARTAN-EHRESMANN THEORY. 1993.
29. LIE-THEORETIC NUMERICAL ANALYSIS, MECHANICS AND DIFFERENTIAL SYSTEMS. 1994,

LIE-THEORETIC ODE NUMERICAL ANALYSIS, MECHANICS AND DIFFERENTIAL SYSTEMS

PREFACE

For the past four years, my work has focussed on two areas: The Lie and Deformation Theory of standard ODE Numerical Analysis and the Differential-Geometric and Lie-Cartan Theoretic aspects of the Generalized Function Algebra Theory of Colombeau, Oberguggenberger and Rosinger. In this Volume I concentrate on the first topic: I hope to get to the second topic in a more systematic way in Vol. 30.

My broad aim on the 'Numerical Analysis of DE's Side' is to develop it as an Extension (or a Deformation!) of the 'Geometric Theory of Differential Systems', as developed by Lie, Goursat, Cartan, Vessiot, Spencer, ...The rest of the Volume deals with more specific parts of what I ultimately hope will be a broad 'Theory of Mechanics Differential Systems, and Their Discrete Approximations'. I concentrate here on ODE's and certain PDE Systems whose 'General Solution' can be found by solving ODE's. On the Applied side, I hope to provide new geometric methodology for thinking about ODE Numerical Analysis, particularly for ODE's which 'live' on general manifolds and those which naturally involve Lie Groups and certain Infinite Lie-Spencer Pseudogroups, such as the Symplectic Pseudogroup which underlies Mechanics. As in previous Volumes, I begin with Polemic, Notes on Lectures, and Introductions.

The Interaction between Numerical Analysis and Mechanics offers a rich and fertile research area. I continue here to study some of the Geometric themes in Mechanics begun originally in my book "Differential Geometry and the Calculus of Variations", and most recently in Volume 27, "Constrained Mechanics and Lie Theory".

TABLE OF CONTENTS

PREFACE

VISTAS INTO APPLICATIONS OF DIFFERENTIAL GEOMETRY   1
AND LIE THEORY IN GENERALIZED FUNCTION AND
DIFFERENTIAL SYSTEM THEORY

LETTERS TO PROFESSORS   13

NOTES FOR LECTURES: THE DEFORMATION AND LIE THEORY   19
OF DYNAMICAL SYSTEM NUMERICAL ANALYSIS

NOTES FOR LECTURES: GENERALIZED FUNCTION ALGEBRAS   29
AND GEOMETRIC DIFFERENTIAL EQUATION THEORY

OUTLINE OF RESEARCH IN GENERALIZED FUNCTION ALGEBRA THEORY   41

1. Introduction.   41
2. An 'axiomatization' for the renormalization and generalized   43
   function approach to the calculus of variations.
3. The vortex equations as the renormalization of the Euler   45
   fluid equations on a Riemannian manifold.
4. The renormalization of the infinite self-energy of   47
   incompressible fluid flows.
5. Furger development of renormalization in the context of the   50
   Cartan differential form approach to the variational calculus
6. Colombeau-Rosinger Theory, the Feynman Quantum Field   51
   Theory renormalizations and 'philosophical' remarks.
7. 1-D generalized functions.   55
8. Regularization of the ϕ^4 quantum field theory and deformation   59
   theory for nonlinear PDE a la Kodaira-Spencer.
   Bibliography   62

I. THE LIE-THEORETIC NUMERICAL ANALYSIS OF SOME FROBENIUS-INTEGRABLE DIFFERENTIAL SYSTEMS

1. Introduction. 71
2. Linear time-invariant ODE's and their Euler approximation. 74
3. Expansions of the Euler approximation of linear ODE's, based on a 'WKB' type of inegral representation of the solutions of linear, time-invariant difference equations. 75
4. More general approximations of the WKB-type for linear time-invariant ODE's. 79
5. The centeredf-difference approximation for linear ODE's and the Cayley transform of the A-matrix.. 80
6. L. Richardson's 'deferred approach to the limit'. 83
7. Systems of difference equations which deform Frobenius Integrable Differential Systems. 86
8. Difference System approximations for linear Frobenius Integrable PDE systems in two independent variables. 88
9. Difference equation approximations to a linear Frobenius System. 90
10. Seperable Frobenius Integrable Linear PDE Systems. 92
11. Frobenius Integrable Linear PDE Systems associated with a smooth family of 2-dimensional Lie subalgebras of **GL(n, R)**. 94
12. Frobenius Integrable Linear Differential Equations and their discrete approximations associated with two dimensionl subalgebras of **GL(n, R)**. 96
13. Difference equatioons obtained from approximation of the General Solution and commutation relations. 98
14. Finite Difference Approximations to Frobenius Integrable Linear PDE's obtained by deformaing the connection and curvature operator of a vector bundle. 100
15. The Frobenius Integrability Conditions for systems of difference equations. 104
16. The nonlinear ODE one-step convergence theorem from the point of view of the theory 105
17. Conditions for the existence of a smooth limitinf dynamical 108

system for deformations of discrete semigroups on manifolds.
18. Final remarks about future research directions. 111

II. COMPUTING THE ORBITS OF LIE GROUP ACTIONS

1. Introduction. 113
2. The geometric and deformation-theoretic setting for some ideas in ODE numerical analysis. 114
3. Towards a Numerical Analysis of Lie group actions on manifolds. 116

III. SOME GEOMETRIC PDE THEORY FOR SMOOTH MAPS

1. Introduction. 121
2. Some concepts from jet space theory. 123
3. Differential Systems and Integrability in the sense of Spencer and Goldschmidt. 130
4. Differential graded algebras and their derivations. 131
5. Moving Coframes of Differential Forms on the Jet Bundles. 135
6. Differential equations, operator and symbols. 138
7. Prolongation of differential operators and differential equations. 140
8. Prolongation of vector fields. 141
9. Prolongation of differential equations. 143
10. General solutions of systems of real-analytic partial differential equations. 145
11. The general solution of the one-dimensional wave equation. 148
12. The Clairault equations: Is "general" and "singular' solution. 150
13. Complete solutions of partial differential equations in the sense of Lagrange and Vessiot. Lagrange-Vessiot foliations and fibrations. 155
14. General solutions that are generated from one particular complete solution. Lagrange-Vessiot fibrations by the method of Lagrange and Charpit. 160
15. The method of Lagrange-Charpit in the context of Cartan's 161

  theory o exterior differential systems.
16. The classical techniques of the "Intermediate Integral"   167
  for Monge-Ampere second-order partial differential
  equations.
17. The families of solutions of the Monge-Ampere Equation   176
  generated by all Intermediate Integrals

    Bibliography   181

IV. THE POISSON-CARTAN ALGEBRA STRUCTURE AND PROLONGATION CONCEPTS ASSOCIATED WITH A SMOOTH, CLOSED 2-FORM ON A SMOOTH MANIFOLD
 1. Introduction.   185
 2. Some properties of the Poisson-Cartan algebra   187
 3. The traditional Hamiltonian Mechanics in terms   189
  of Poisson-Cartan structures.
 4. Cauchy characteristic vector fields and curves   192
  associated with a closed 2-form.
 5. The Cauchy Characteristic foliation associated with a   194
  closed 2-form.
 6. Mechanical systems and their symmetries in terms   196
  of Poisson-Cartan structures.
 7. Critical points of functions whose Poisson Bracket   199
  with the Hamiltonian vanishes and orbits of
  one-parameter symmetry transformation groups which
  are also extremals of the mechanical system.

    Bibliography for Mechanics   200

V. REGULARIZATION OF SINGULARITIES OF ORDINARY
  DIFFERENTIAL SYSTEMS

 1. Prolongations of ODE systems.   207
 2. ODE systems and their prolongations defined by systems   208
  of first-degree differential forms, i.e. Pfaffian Systems.
 3. A method for Regularization of 1-D Hamiltonian Systems.   211

    4. The 1-D Regularization Algorithms in terms of a   215
       Moving Frame for the state space X.
    5. Regularization of a 1-D Newtonian particle.   218
    6. Reformulation of the 1-D Newtonian regularization   221
       in terms of symplectic structures and fiber bundles.
    7. Final Remarks   223

        Bibliography   224

VI. SOME LIE-THEORETIC ASPECTS OF THE 2-BODY PROBLEM OF
    CELESTIAL MECHANICS

    1. Introduction.   225
    2. The 2-Body Problem of Newtonian Celestial Mechanics   226
       and the Runge-Lenz Vector.
    3. The description of the Kepler Hamiltonian and   228
       its symmetries in terms of finite dimensional
       Lie algebras and Lie pseudogroups.
    4. The commutation relations on the energy surfaces.   231
       'Deformation' of the Lie algebras defined by the
       Runge-Lenz vectors as a smooth function of the energy.
    5. The pseudogroup action of SO(4, R) on the positive-energy   236
       part of the phase space regularization of the
       Kepler trajectories.

        Bibliography   240

VII. CARTAN'S FORMULA FOR THE POISSON BRACKET
     GENERALIZED TO CONSTRAINED SYSTEMS.

    1. Introduction.   241
    2. Symplectic manifolds, Poisson Bracket and   241
       symmetries of vector field systems generated
       by infinitesimal symplectic automomorphisms.
    3. The Poisson Bracket operation on smooth real-valued   243
       functions on a symplectic manifold.

4. The classical formula for Poisson Bracket.	244
5. Cartan's formula for the Poisson Bracket in terms of exterior multiplication.	246
6. A generalization of the Poisson Bracket for constrained systems using Cartan's formula.	250
7. The Poisson-Cartan Bracket on symplectic submanifolds of symplectic manifolds.	251

VIII. LAGRANGE'S EQUATIONS AND EHRESMANN CONNECTIONS IN THE TANGENT BUNDLE

1. Introduction.	255
2. Lagrange's unconstrained equations in local coordinates.	257
3. Lagrange's unconstrained equations as the differential equations for straight lines in an Ehresmann connection for the tangent bundle.	259
4. The differential equations in local coordinates for the straight lines of an Ehresmann connection in the tangent bundle and their equivalence to those of a Lagrangian mechanical system.	261

IX. EHRESMANN CONNECTIONS IN DIFFERENTIABLE FIBER SPACES AND CONSTRAINED MECHANICAL SYSTEMS

1. Introduction.	265
2. Differentiable fiber spaces and the foliation by vertical vectors.	267
3. Ehresmann connections and horizontal liftngs of curves and vector fields in the base.	268
4. Curvature for Ehresmann connections.	271
5. The horizontal lifting of curves and vector fields relative to an Ehresmann connection and parallel transport.	272
6. Horizontal completeness.	277
7. Path systems defined by Ehresmann connections on sub-bundles of the tangent bundle.	279
8. The constrained Lagrange Equations in Local Coordinates,	282

in the case that the constraint subbundle is of codimension one.

VISTAS INTO APPLICATIONS OF DIFFERENTIAL GEOMETRY AND LIE THEORY IN GENERALIZED FUNCTION AND DIFFERENTIAL SYSTEM THEORY

My plan for this Volume is to assume known some of the basic analytical facts about ordinary differential equation numerical analysis and to attempt to impose further geometric and algebraic structure on them. I believe that much the same program can be carried out as well for partial differential equations. (I hope to develop this in a later Volume.) As I have emphasized throughout my work on mathematical physics, control theory and applied mathematics, the Theory of Deformation of Algebraic and Geometric Structures plays an important, but usually unrecognized, mathematical role. In Volume 25, I have pushed this insight even harder and suggested that some sort of 'deformation theory' would play a role in Computer Programming Theory.

*Despite the success (ever since the work of Poincare!) of geometric insight in the study of **differential** equations, the development of the analogous side of **difference** equations (and of course Numerical Analysis has a good deal to do with the 'approximation' of differential equations by difference equations!) has seriously lagged. In recent years, the Lie Theory (i.e. the study of the relations between 'symmetries' and 'solutions') of Differential Equations has come into its own in both Pure and Applied contexts. I am proud to say that I started to work in this area at the time of my Ph. D thesis, in the mid-1950's. My 1968 book "Differential Geometry and the Calculus of Variations" (whose first versions appeared as Lincoln Lab and UC Berkeley notes in the early 1960's) contains what I believe is the first treatment in a modern context of some of Lie's material. Despite all the emphasis that "Computational Mathematics" has received in recent years, little has been done to extend Lie's ideas to Difference Equations. Starting in this direction was one of my main goals in developing the machinery in Volume 25. From Ehresmann I learned that one approach to understanding and generalizing Lie was to 'translate' the 'geometric' background of Lie into the algebraic language of Category*

Theory. *Of course, as always with Category Theory, there is the danger of outrunning one's source of 'intuition' in terms of the material from which one is abstracting the category-theoretic heart and ending up with a collection of Definitions. I admit that this probably happened to me in Vol. 25! The material presented here on the Lie and Deformation background to Numerical Analysis will restore some balance! What I hope to see done (and I have certainly not finished the job here) is to systematize and 'geometrize' the study of the Numerical Analysis of ODE's, and also to prepare the geometric and algebraic ground for generalization to PDE's.*

I have been privileged to witness the introduction of many new and exciting ideas in mathematics, physics and engineering. One such period was my graduate student years (in 1952-55) in Princeton, Paris, Strasbourg and Amsterdam. New methods of differential geometry and Lie group theory were introduced; sheaf and fiber bundle theory and associated homology and homotopy theory were developed; and my thesis advisor Don Spencer began his work recasting the analytical and algebraic theory of differential systems.

A second such period was my brief (1959-61) experience as a Staff Mathematician at Lincoln Laboratory of MIT. In contrast to the first period, this was an exciting time for the development of Applied Mathematics. It was the beginning of the Space Program, and a new generation of engineers were beginning to tackle problems which had not yet been modernized, emphasizing the use of computers, methods of signal processing, control and filter theory, etc. In addition, there was a new and improved collection of mathematical tools -probability and stochastic theory, linear algebra, dynamical systems, etc. - available to think about the engineering problems in more mathematical terms. I was on the scene to suggest application of differential-geometric methods to handle the sort of nonlinear problems which were on the horizon.

My third experience came in 1962-69 with the application of Lie group theory and differential geometry to elementary particle physics. I was among the first to prepare the way (in many books and papers) for the renewed interaction between 'Fundamental' Physics and Differential Geometry which occurred in the 1970's.

I believe that in two fields we are on the brink of a new burst of such a creative research breakthough; this time we may gain fundamental new insights into **nonlinear physics** using the new Colombeau-Rosinger-Oberguggenberger insights and on into Computational Mathematics, developing ways of thinking about 'computation' which parallel the 'algebraic' ways that we think about Geometry. I have been thinking about the C-O-R theory from the 'geometric analysis' point of view for several years. Partial results have been presented in Volumes 26, 27 and 28. I hope to get to a more definitive version of my ideas in Volume 30; this Volume concentrates on the Numerical Analysis and Deformation side of the theory, except for several Notes for Lecture in the beginning of the book. As in much of my 'applied' work over the past thirty years, I also see differential-geometric methodology as the basis for unification and further development.

The Numerical Analysis of Ordinary Differential Equations (ODE's) has developed as a 'pragmatic' computational theory. Since it intrinsically involves 'nonlinear' mathematics, it has not been feasible to treat it with the general tools of Functional Analysis, as have other branches of Numerical Analysis. My aim is to recast it in terms of manifold, Lie pseudogroup and topological dynamics theory, providing material which complements that in the standard treatises,e.g. that by Gear and Bulirisch-Stoer. For example, I will not attempt to reprove the standard analytic material on the asymptotic nature of the Numerical Analysis ODE algorithms.

In Lie-theoretic terms, the theory of ODE's involves smooth continuous-paramter one-parameter pseudogroups of transformations on manifolds. One point of view in this book is that Numerical Analysis of ODE's involves approximation and deformations of such pseudogroups in terms of discrete-parameter pseudogroups. Here is one way of thinking about such deformations.

Suppose given the following data:

X is a smooth manifold (1.1)

$\{t \rightarrow g(t): X \rightarrow X\}$ is a continuous one-parameter pseudogroup of diffeomorphisms of X. The trajectories of the ODE are the orbit curves: (1.2)

$$\{t \rightarrow x(t) = g(t)(x(0))\} \quad (1.3)$$

Numerical Analysis now provides the following data:

An interval **H** of real numbers containing 0. (1.4)

A smooth map: **H** $\rightarrow$ Local diffeomorphisms of X
$\{h \rightarrow \phi(h): X \rightarrow X\}$ (1.5)

Definition. $\{h \rightarrow \phi(h)\}$ is a **one-step numerical analysis approximation** to the orbit 1.3 of pseudogroup 1.2 if the following condition is satisfied:

$$g(t)(x(0)) = \lim_{h \to 0} [\phi(h)]^{t/h}(x(0)) \quad (1.6)$$

One can go further and associate notions of **order** with such approximations. Such material for Euclidean spaces is standard in ODE numerical analysis books. I hope to develop it - and a 'geometric' theory of Asymptotic Expansions - for curves and other maps on manifolds in Volume 30. I consider this a part of 'geometry', an extension of the classical notions of 'orders of contact' which underly the Ehresmann jet-bundle formalism. For example, one can try to define the concept of:

'Two maps are asymptotically equal to such-and-such an order at a point' in terms of Riemannian geometry.

Linear ODE's (particularly the constant-coefficient kind) have the happy property of being 'explictly soluble' in terms of a known (and interesting) algebraic structure (i.e. Linear Algebra) and form a class whose asymptotic analysis reflects a good deal of the structure of many non-linear ODE's. Accordingly, I will emphasize the Example of the Numerical Analysis of systems of linear ODE's.

Another general question is the following one:

> If the given ODE belongs to a Lie algebra of vector fields
> i.e. the solutions belong to a given Lie pseudogroup,
> when can Numerical Analysis algorithms be found to a given asymptotic order which also generate solutions lying in the pseudogroup?

(This subject has been popular in recent years for the 'symplectic pseudogroup' case, leading to the study of 'symplectic integrators'.)

In Volume 21 of "Interdisciplinary Mathematics" I have suggested that some of the algorithms of Linear Algebra Numerical Analysis are closely linked to certain topics in Lie Group Theory. (This work has been pursued further by Clyde Martin and Greg Ammar.) In this work. I will proceed in a similiar spirit to suggest that some of the Algorithms of ODE Numerical Analysis may be usefully considered from the point of view of Lie Pseudogroup and Deformation Theory.

In Volume 25, I tried to sketch my Vision of a Grand Unification of Recursive-Function Based Computer Science, "Geometric" Logic, systems of differential equations, etc. (Throughout my career, I have enjoyed sketching out this 'Big Picture" rather than becoming extensively involved with spelling out all the details!) Admittedly, I was not very successful in filling in the Details of this Vision: I hope that the material to be developed in this Volume will serve to inspire other researchers to take up this task of Grand Unification!

Deformation Theory, as developed in the pure mathematics literature primarily in the 1960's has also been a great inspiration in much of my 'applied' work. In this Volume I hope to continue this direction, emphasizing the role of deformation-theoretic ideas in ODE and (to a limited extent!) Differential System Numerical Analysis. From my point of view, Deformation Theory enters very naturally in all sorts of Applied situations because the Applied work inevitably involve such objects as diffential or difference equations **depending on parameters.** The situation often involves introducing some sort of equivalence relation on such systems and their solutions; this leads to typical geometric equivalence problems, of the sort considered originally by Riemann, Lie and Cartan. Deformation Theory was introduced into the mathematical world with the work of Spencer and Kodaira in the late 1950's and early1960's. As a student of Spencer, when I started to look at Physics in the 1960's I saw his deformation-theoretic ideas as a unifying mathematical methodology for much of the mathematical jumble which is inherent in the structure of the physical sciences and associated engineering disciplines.

A basic idea of Generalized Function Algebra Theory is to define the 'generalized' solutions of differential equations by means of families of 'ordinary' solutions which 'depend on parameters'. This also indicates that a fundamental mathematical connection between Generalized Function Algebra Theory and the Geometric Theory of Deformations, as initiated in the work of Spencer and Kodaira. I have already indicated such a connection in Volume 26, in the context of Quantum Field Theory. As I hope to demonstrate in this and future Volumes, much the same thing happens for the Numerical Analysis and Computer Science description of physical phenomena governed by differential equations. However, the researchers in Numerical Analysis have concentrated on the 'practical' side of their beloved 'algorithms', and have not (as far as I can tell!) asked themselves the sort of questions Geometers have been used to for over one hundred years:

What is the Mathematical Structure that the Algorithm represents? How do the 'practical' and 'complexity' aspects reflect this Structure?

How does one set up Equivalence Relations on the class of Algorithms so that two such are equivalent if and only if the 'reality' they describe is the same?

Remark. Notice that much the same conceptual progress has taken the years since about 1950 in **Generalized Function Theory**. Sobolev, Schwartz, and their cohorts concentrated on the 'algorithms' for defining 'Generalized Functions' derived from studying **linear** PDE, noticed that these Algorithms were unsatisfactory for **nonlinear** phenomena, and gave up at that point! The great conceptual progress in the work of Rosinger and Colombeau (since the late 1960's in Rosinger's case) was precisely to put the question in the same framework, i.e. to study **equivalence classes** of 'algorithms'. Unfortunately the work and ideas of both mathematicians was not absorbed (as it still is not to this day!) by 'mainstream' mathematicians, who should have known know better: Think of 'Dedekind Cuts'!

My aim in developing Generalized Function Algebra Theory in geometric directions is to prepare the ground for Applications to those parts of the physical sciences which involve non-linear partial differential equations, particularly when the desired solutions have singular behavior. As I have explained already in Vols. 26 and 27, Quantum Field Theory and Fluid/Vortex Dynamics are the prototypes of such theories. Now, one of the most important topics in the theory of non-linear partial differential equations is the study of **conservation laws** and **symmetries.** The geometric theory of differential systems and PDE's (begun systematically by Lie and Cartan) involves two basic geometric objects:

> Lie Algebras of Vector Fields which are Infinitesimal Symmetries
>
> Differential Form Algebras which define Conservation Laws for the system.

The essence of 'Hamitonianness' is a sort of duality between these two Geometric Structures. These structures are especially important for the theory of the Generalized Function Algebra solutions, for reasons which I will go into in more detail in the text. Here is one immediate comment:

> A Conservation Law (which is just a differential form!) for a PDE defines, by integration, a real-valued function on the space of 'regular' solutions of that PDE. However, this function develops singularities when it is extended to the 'generalized' solutions. Eliminating this singularity is what is involved in Regularization and Renormalization. After Regularization the action of the Symmetries on solutions is (possibly) altered. The quantum field physicists have encountered this phenomenon under the slogan 'Anomalies'. I was one of the first (in the 1960's) to study such phenomena from a Geometric point of view: See my book "Lie Algebras and Quantum Mechanics".

I also have an ulterior motive in linking Numerical Analysis of Differential Systems and the Generalized Function analysis of Differential Systems. Both involve something like a 'deformation' of a 'smooth' system; I expect that someday there will be a unified way of looking at such things.

Another topic to be initiatiated and treated in this Volume and expanded in the succeeding ones) is the generalization of the **Poisson Bracket** operation that is so important in classical and quantum mechanics. (This material may be also be thought of as an addendum to that on Constrained Mechanics in Volume 27). Generalizations are desirable in many directions. Here is a possible list.

   a). A Poisson Bracket associated with constrained 1-D mechanics problems, 'holonomic' and 'non-holonomic'.

   b). A Poisson Bracket associated with constrained and singular 1-D variational problems.

   c). A Poisson Bracket associated with field-theoretic mechanics and variational problems.

d). A Poisson Bracket associated with what is now called a 'Poisson Structure' (and that I called a 'cosymplectic structure' in some of my books of the 1970's). It is defined by a contravariant tensor field, which is its 'symbol' as a differential operator.

e). A Poisson Bracket associated with quantum particle and field-theoretic physics situations.

Again, there is much scattered through my own work on material pointing in these directions. In this Volume, I will begin a direction of unification based on ideas developed by Elie Cartan and expounded by him in his 1920's book "Lecons sur les Invariants Integraux". (The basic ideas are already in a paper he publishd in the 1890's!) As in all of Cartan's geometric and physical work, his version of Poisson, Symplectic, etc, is based on the differential form calculus. Here is a brief description of a setting for Cartan's ideas.

Start of with the classical situation, and the following data:

$X = R^{2n} = \{x = (p, q)\}$; $p = (p_1, ..., p_n)$; $q = (q^1, ..., q^n)$

Indices: $1 \leq i, j, ... \leq n$

$\{f, f'\} = \Sigma_i [\partial f/\partial p_i][\partial f'/\partial q^i] - [\partial f'/\partial p_i][\partial f/\partial q^i]$

f, f' are real-valued, smooth functions on R^{2n}.

$(f, f') \rightarrow \{f, f'\}$ is a R- bilinear, first-order differential operator.

It is now well-known how to express the Poissson bracket formula in terms of the usual symplectic structure defined by the folowing 2-form:

$\omega = \Sigma_i dp_i \wedge dq^i$

on R^{2n}. Usually, the Poisson-Bracket formula is defined by means of a Lie derivative operation. Cartan now remarks that the classical formula can be defined completely in terms of the exterior derivative operation 'd' and exterior multiplication operation '∧' as follows:

$$df \wedge df' \wedge [\omega]^{n-1} = (\text{constant}) \times \{f, f'\}[\omega]^n$$

where:

$[\omega]^n$ = the exterior product of n copies of the 2-form ω.

Cartan's formula is now readily generalizable in many directions. Here is one of the simplest:

> Suppose that, in addition to the above data, one is given another 2-form θ. Given functions (f, f') on X, define:
>
> $\{f, f'\}_\theta$
>
> by the following formula:
>
> $$df \wedge df' \wedge \theta \wedge [\omega]^{n-2} = \{f, f'\}_\theta [\omega]^n$$

Then, if 'θ' represents 'costraints', the operation:

$$(f, f) \longrightarrow \{f, f'\}_\theta$$

defines a **Constrained Poisson Bracket Operation.** My ultimate intention is to study such geometric objects and their 'applied' origins and ramifications in more detail.

Remark. Cartan's Constrained Poisson Bracket Operation not only subsumes the well-known "Dirac Bracket" but it long pre-dates it!

One broad mathematical theme in the ideas associated with 'Poisson Bracket' is the assignment of 'algebra' to 'geometry' and 'physics'. The situation on the ODE Numerical Analysis part of my work in this volume is similiar: I want to understand the geometric and algebaric setiing of the classical ODE numerical algorithms. The way I do this is to associate these algorithms with the algebra deformation theories developed in the period 1955-65 by Frohlicher, Gerstenhaber, Nijenhuis and R. Richardson, and used by me to study what I called 'Analytic Continuation' of group representations. I hope to get back to the study of these topics in more detail in later volumes.

Finally, I introduce the study of the Geometric Foundations of the Theory of Regularization of the singularities of the ordinary differential equations of mechanics, as initiated in the work of Euler, Levi-Civita, Kustaanheimo and Stiefel. My long-term goal, which is only initiated here, is the embedding of this work in the Theory of Geometric Structures.

LETTERS TO PROFESSORS

For most of my career, I have been in the No Man's Land between the 'pure' and 'applied' worlds. Since I believed that there was a middle way, I coined the slogan 'Interdisciplinary Mathematics' to denote what I do! It has been a long and lonely road since I began at Lincoln Labs in the late 1950's. To express my frustration I (like Herzog, in the novel by Saul Bellow) **write letters:** *Here are some recent samples.*

LETTER A

In the issue of Dec. 1993, the *AMS Notices* has published an exchange between Profs. R. Gangoli and S. MacLane concerning their ideas for the orientation of that part of the US mathematical community centered around the AMS, particularly the relative weight to be assigned to Education and Research. While there is much that I agree with in both presentations, I feel compelled to point out (as I have done in many places for the past thirty years!) that neither piece seems to have adequately taken into account the renewed interaction between Mathematics, Science and Technology which has taken place in our professional lifetimes, particularly in the past twenty years.

Consider such terms of popular and scientific discourse as:

Artificial Intelligence, Black Holes, Cellular Automata, Chaos, Computer Languages, Computer Programs, Control, Filtering, Fractals, Gauge Fields, Modelling, Neural Nets, Quarks, Signal Processing, Supersymmetry, Symmetry, Wavelets, ...

Each of the areas represented by these terms is based on a part of 20th century science which was originated by professional mathematicians. With such a record of accomplishment, why are we in the position of begging Congress - and the NSF/DOD research bureaucracy! - for Pennies while other areas of science and technology squabble over Millions, if not Billions? Congress quite rightly wants the National Science Foundation (and other agencies which support science and technology) to emphasize work which will strengthen the national economy and educational structure: Should not Mathematics be at the head of the line? Why is not someone in the AMS pointing this out loud and clear to the media, the general public and Congress? At the minimum, what is badly needed is some explicit encouragement for the sort of thing I described above and a clearing of the bureaucratic hurdles.

As examples of what I mean by "Interdisciplinary Mathematics", let me cite two interests of mine: The first is the work of Gerry Sussman and Jack Wisdom on the numerics of the long-time behavior of the planets, the second the C. O. R. Generalized Function work and its ramifications into Geometric Analysis. Sussman and Wisdom's methods involve a combination of clever computer design and approximation technique. It's clear to me that this involves some new areas of Deformation Theory, since what are being 'deformed' are basically groups, and, perhaps, Hopf Algebras. I'm trying to work out as many of the ramifications of this (e.g. the relations between group deformations and Eilenberg-MacLane cohomology are only partially done, and that in a very limited way, in the literature that I know about) as I can, but its slow going.

Second, I have been working in the directions pioneered by Colombeau, Oberguggenberger and Rosinger on the old 'problem' of Multiplying Distributions. They have done a terrific job (starting over twenty years ago!) combining techniques of algebra, analysis, and just plain stick-with-itness and mathematical inventiveness, for what is obviously an extremely important area with many applied ramifications (fluid vortices and turbulence, shock-waves, interacting quantum field theory, non-smooth control theory, etc.). No one (aside from me; I am adding my knowledge of Lie and Differential Systems Theory) on this side of the Atlantic has taken serious notice of this, and again I have applied to NSF for 3 years without success so far (and zip from the Mission boys!)

LETTER B

I recently heard of some statements in connection with material on Geometric PDE Theory to which I must take exception! As I understand them, the statements go something like this:

(a) In order to be considered as 'applied mathematics' material must be in notations that 'applied mathematicians' can understand, and:

(b) Before introducing 'abstract' notations a geometer should show an example for which it is superior.

Now, I realize that you were not trained in differential geometry and presumably know little of the history of the subject. The First Law of Differential Geoemetry (due to Nijenhuis) is that:

Differential Geometry is the Study of Invariance Under Change of Notation.

Because of the way the subject has developed historically, this Law remains valid today (as it does not in Analysis and Algebra). The part of Differential Geometry involved in the book you were discussing originated in the work of Lie, Darboux, Goursat, Cartan, Vessiot, Ehresmann, Spencer, ... These gentlemen introduced truly brilliant **ideas** about the way to handle Differential Systems, without the tedium and detail of the 'classical' methods which you insist 'applied' mathematicians must learn. Now, the 'classical' methods do indeed have the great virtue of being relatively explicit and concrete, but they often lead to what Cartan (paraphrasing!) might call 'un debauch de formules', which hides the true 'geometric' essence of what is going on. See also the statements in the Introduction of Ricci and Levi-Civita's classic 1900 paper on Tensor Analysis about the need for what the computer scientists would call a 'higher-level language' for the description of complicated and messy analytic formalisms!

It also seems to me that a) is condescending to Applied Mathematicans, implying (for example) that they are too dense to learn new notations even if they see the motivation and applicability. Must of my 'applied' work has been with physicists: I have found that they will learn **any** notations (and indeed 'invent' them themselves after they understand a piece of what we mathematicians have already done!) if the relevance to what they are doing is pointed out. Of course, as a compulsive writer of semi-expository things, I know what you **might** mean, e.g.. that something is a hell-of- alot- clearer if it can be explained in notations which are common knowledge across a wide spectrum of interests, but there is an anti-intellectual and anti-scholarly flavor to the way you put it.

As for b), presumably that is a dig at jets, category-theoretic ideas, etc. Now, God Knows that Ehresmann's work is difficult to understand and his notations are difficult to work with, but what does not seem to be common knowledge is that his 'abstractions' are taken directly from Lie! (See the Translations of Lie which I have published.) It is also a mathematical language which is necessary to understand what Cartan said about 'Infinite Lie Groups', which is, in my humble opinion, still the most important seminal work in Differential Geometry in the 20th century! (And as a student of Ehresmann, let me assure you that 'understanding Cartan' was in the forefront of his mind!) Again, its our job as Scholars to put this stuff in a form for which it applies to specific problems, but you (and the other bitchers and moaners about jets, categories, etc.: I have done my own share of such bitching!) are basically complaining about the two geniuses who originated the subject, Lie and Cartan. Ehresmann was instrumental in putting their work into a form where it could be assimilated into the rest of Mathematics.

LETTER C

As a bit of Emperical Research into the background of the recent literature published in the *Notices of the AMS* about the employment situation, the clout of the mathematics profession in the national science structure, etc., I consulted the **Catalog of Research Opportunities in US Government Laboratories,** published by the National Research

Council of the National Academy of Sciences, the National Academy of Engineering and the Institute of Medicine. There is **one** page devoted to Mathematics (in a total of **178**!), listing the following 15 topics:

Adjustment of Risk Estimates for Exposure Measurement Error; Applied Mathematics and Mathematical Modelling; Artificial Intelligence, Chemical Exposure Assessment Methodology; Computational Fluid Mechanics and Heat Transfer; Epidemilogic Studies of Occupational Diseases; Epidemilogic Studies of Workers Exposed to Dioxin; Formal Methods; Human Computer Interaction Research; Mathematical Optimization Modelling; Matrix Computation for Structured Problems; Modelling of Multidimensional Flow in a Liquid Propellant Gun; Nonlinear Dynamics and Stochastic Modelling; Occupational Health Hazard Evaluations; and Statistical Inference Based on Replicates.

May I suggest that the eminent mathematicians who have been honored by membership in the National Academy of Sciences make a greater effort to enlighten their colleagues in the National Research Council about the full spectrum of advances which have been made by the mathematics profession world-wide, and the opportunities that this progress presents for support of the national effort in science and technology?

This is important less for its intellectual implications than for the opportunities which might be opened up for under or un-employed mathematicians. Such an expanded list might also be useful in persuading American students who are thinking of studying mathematics that there might, someday, be job opportunities as mathematicians beyond temporary teaching positions!

LETTER D

Much, much more could be done to impress the politicians, indeed to throw their own rhetoric back in their faces, since we have much more to contribute (at least proportional to the Bucks involved!) to National Competitivness than the other 'pure' sciences. (For example Kostant tells me that Buckyballs involve some new exotic group theory!).

Working with Ford, GM, etc. is not at all hopeless and fruitless. Academic, NSF-supported (and AMS-backed) mathematicians could give lectures to engineers and scientists on their own work and how it might be useful. This would inevitably lead to consulting, jobs for students, post-docs, etc. If nothing else, there is just a tremendous amount of Real Mathematics being thrown up by the advances in computer and communications technology, etc. I am skeptical that setting up a new academic bureaucracy on 'Industrial Mathematics' will do very much beyond inflating a few academic egos. What is needed is getting some mathematically creative people out into the world and environments where things are going on.

As an outsider getting his information mainly from the NY Times and Science I have the impression that much of our difficulties in getting 'our' Little Science point of view over to Congress and the Politicians can be traced back to the dominance of NSF by physicists and their Big Science mentality. If I was a congessman hearing all that Bull over the years about **'Funding SSC in Order to Cure Cancer and Discover God'** I too would revolt and demand that **'By God, Those Guys in NSF Produce Something That Will Benefit Us Taxpayers'**!

Can nothing be done at some collective/AMS/SIAM/MAA level both to make such propaganda in a politically effective way and to do something substantive and practical about changing the financial and intellectual environment in which such things are decided?

THE DEFORMATION AND LIE THEORY OF DYNAMICAL SYSTEM NUMERICAL ANALYSIS

A talk for The Workshop on "Application Specific Techniques in High Performance Computing Environments" at The Fields Institute

Abstract

"Computer Science" must relate the Worlds of Languages and Discrete Event Processes to the mathematico-physical world of Dynamical Systems, Differential Equations, Lie Groups, etc. Involves a Deformation of Algebraic Structures. As a 'workshop' the subject matter of the Numerical Analysis of Differential Equations is ideal. In this Lecture, I will describe in broad terms some 'geometric' ideas in 'support' of the work of Abelson, Sussman and Wisdom reported in last week's session. I will emphasize broad mathematical ideas rather than the usual Numerical Analysis nitty-gritty. On the Control side, an aim is also to develop what I call Trajectory Control, i.e. a Control Theory whose elements are based on mathematical objects like curves.

For example, in Typography (as interpreted 'geometrically' by Mathew Halfant) one considers a letter of the alphabet as a collection of points in R^2, i.e. an element of an R^n which is changed via a vector field in R^n and interpolated via splines to define the figure.

Brief explanation of 'where I am coming from'.

I started - in the mid-80's - talking with George Meyer and Gerry Sussman about a synthesis of Computer Science and Control ideas in order to build a mathematical machinery better adapted (e.g. than Discrete Event System Theory) for describing the Continuous-Differential Equations Dominated World of the Physical Scientist/Engineer in terms of Computer Science and Logic. In Mathematical terms, Deformation Theory (e.g. of semigroups, categrories, etc., of a type which does not yet exist!) gives and intellectual framework for doing this.

Wrote "Geometric Computing Science: First Steps", i.e. "Interdisciplinary Mathematics", vol. 25. Admittedly very incomplete and unsatisfying. Basic Mathematical idea is that Category Theory is the appropriate framework for describing Logic, Languages, Dynamical and Recursive Systems in a way that might lead to a satisfactory theory of their 'interaction'.

It is my "Erlangen Programm". Like Felix Klein's (written in 1870's) it might take 80 years to work out! An appropriate "Deformation Theory" should relate the 'discrete' world of CS and the 'continuous' world of DE's.

Need to understand many Examples and Theories in much greater depth! I turned to ODE Numerical Analysis to find such Examples.

Here is an example of the sort of 'structure' I want to impose.

A **Dynamical System in the Mathematical Sense** is a smooth action:

$$\{t \longrightarrow \phi(t): X \longrightarrow X\}$$

of the semigroup R^+ (the additive group of non-negative real numbers) on a manifold X.

A **Dynamical System in the Computer Science (and 'Numerical Analysis' Sense)** is the mathematical structure given by the following data:

A): Manifolds X, Y and H.

   N denotes the non-negative integers.

B). A submersion map $\pi: Y \to X$.

C) A smooth map $\{(n, h) \to \psi(n, h): Y \to Y\}$ of:

   N x H --> {smooth maps of Y into itself}

such that:

   The map $\pi\psi$ can be used asymptotically to calculate (to a certain order) the orbit curves $\{t \to \phi(t)(x_0)\}$ of a 'mathematical' dynamical system on X.

Various examples and ramifications of this new geometric structure will be in a forthcoming book. Here, will describe some traditional ODE Numerical Analysis lore in a form better adapted to this general framework.

An algebraic version of one-step methods

$\{t \to \phi(t): X \to X: t \in R\}$ a one parameter group of diffeomorphismss of maniold X.

V = infinitesimal generator

V(x) = tangent vector to curve $t \to \phi(t)(x)$ at t=0.

$\{x(t) = \phi(t)(x_0)$ = orbit curve = solution of ODE

Approximate solutions. Given:

$\{h \to \alpha(h): X \to X\}$

$y(t, n) = \alpha(h)^n(x_0); h = t/n$

One-step, first order convergence Theorem:

> If: tangent vector of $\{h \longrightarrow \alpha(h)(x_0)\}$ at $h = 0$
>
> $= V(x_0)$
>
> $\lim_{n \to \infty} y(t, n) = x(t)$

Deformation-Theoretic meaning:

> Semigroup $\{n \longrightarrow \alpha(h)\}$ deforms 'to first order' to semigroup $\{t \longrightarrow \phi(t)\}$

In ODE Numerical Analysis, one deals with the 'infinite order jet' of mapping $\{h \longrightarrow y(t, t/n)$.

However, we need a more systematic Deformation Theory! Possibly, via Hopf Algebra approach?

An Algebraic Structure for the Richardson Method

Richardson's Method, in its classical form, can be described as follows Suppose given the following data:

> A representation of a real-valued function $\{h \longrightarrow F(h)\}$ of a real variable h of the following form:
>
> $F(h) = a_0 + a_0 h^p + G(h) h^q$
>
> Let λ be a real number such that: $0 < \lambda < 1$. Set:
>
> $F_1(h) = F(h) + [F(h) - F(\lambda h)][\lambda^p - 1]^{-1}$

Then, as h --> 0, $F_1(h)$ is a better approximation to a_0 than $F(h)$.

Iterating the construction, we obtain a sequence:

$$\{F(h), F_1(h), ... \}$$

of better and better approximations to a_0.

This method for improving approximation has many variations. Its widest use seems to be in constructing algorithms for ODE numerical solutions.

Richardson's algorithm in terms of mapping into real vector spaces.

V = a real vector space

H = an interval of positive real numbers with 0 as left end-point.

$\{\phi, \phi_1, ..., \phi_n\}$ is a sequence of maps: $H \rightarrow V$.

$\{\alpha, \alpha_1, ..., \alpha_n\}$ is a sequence of linear maps: $V \rightarrow V$.

Set:

$$F(h) = \alpha(\phi(h)) + ... + \alpha_n(\phi_n(h))$$

The Question:

How can $\{\alpha, \alpha_1, ..., \alpha_n\}$ be chosen so that $F(h)$ is best approximation to $F(0)$ as $h \rightarrow 0$?

Power series in h, etc.

Special case:

$$\phi_j(h) = \phi(\lambda_j h), \quad 0 < \lambda_j < 1$$

Non-linear Richardson Algorithms.

For example: via Bulirisch-Stoer Rational Funcion Interpolation.

$$F(h) = \rho(\phi(h), ..., \phi_n(h))$$

ρ is a map which is determined, e.g.
by suitable Interpolation properties.

Problem: to make a satisfactory algebraic deformation theory about all of this! Deformations of semigroups and Hopf Algebras?

Example: Linear ODE

A is nxn real matrix

Approximations to $\{t \rightarrow \exp(tA)\}$

Euler Approximation:

$$\{n \rightarrow (1 + hA)^n : t = nh\}$$

$$(1 + hA)^n = \exp(n \log(1 + hA))$$

$$\exp(tA) = \lim_{h \rightarrow 0} (1 + htA)^n$$

$y(t, h) = [\log(1 + Ath)]/h = \log[(1 + Ath)^{1/h}]$

$= \int_{[0, 1]} [tA/(1+ tAhx)]dx$

$= \Sigma_j(-1)^j[tA]^{j+1}h^j/(j+1)$

ASYMPTOTICS ON MANIFOLDS

The Numerical Analysis of ODE's involves Asymptotic Analysis.

The extension of the 'standard theory' to dynamical systems on manifolds. Standard theory' builds on classical notions of 'asymptotic expansions in terms of power series';

These concepts not covariant under diffeomorphism, hence do not extend routinely to manifolds.

Given the following data:

A manifold X.

A curve $\{t \to x(t): 0 \le t \le a\}$ in X.

An interval $\mathbf{H} = \{h: -b \le h \le b\}$ of real numbers.

A map: $\{(t, h) \to \aleph(t, h)\}$ of $[0, a] \times [-b, b] \to X$.

We want to give a more precise meaning to the following statement:

For given t, $\aleph(t, h)$ approximates $x(t)$ to a given order as $h \to 0$.

It requires only a topological structure for X to make invariant sense out of the statement that:

$$\lim_{h \to 0} \aleph(t, h) = x(t)$$

One way to extend condition is to require that:

X has a metric space structure:

$$d: X \times X \longrightarrow R$$
$$(x, x') \longrightarrow d(x, x')$$

such that the following condition is satisfied:

The topological structure associated with the metric agrees with that associated with the given manifold structure on X.

Definition. For given real numbers t and $0 \leq r$, $\{\aleph(t, h)\}$ approaches $\{x(t)\}$ to order r iff:

$$\lim_{h \to 0} \left[d(\aleph(t, h), x(t)) / h^r \right] = 0$$

Richardson's 'deferred approach to the limit'.

Suppose that:

$$X = R = \text{real numbers.}$$

$$\aleph(t, h) = x(t) + \aleph_1(t)h + \aleph_2(t)h^2 + \ldots$$

Let:

h_1, h_2 be two arbitrary real numbers,

Then:

Ideas in Numerical Analysis

$$ℵ(t, h_1) = x(t) + ℵ_1(t)h_1 + ℵ_2(t)h_1^2 + ...$$

$$ℵ(t, h_2) = x(t) + ℵ_1(t)h_2 + ℵ_2(t)h_2^2 + ...$$

$$h_2 ℵ(t, h_1) - h_1 ℵ(t, h_2) = x(t)(h_2 - h_1) + ℵ_2(t)[h_2 h_1^2 - h_1 h_2^2] ...$$

hence:

$$[h_2 ℵ(t, h_1) - h_1 ℵ(t, h_2)]/(h_2 - h_1) = x(t) + ℵ_2(t)[h_2 h_1^2 - h_1 h_2^2]/(h_2 - h_1)...$$

This identity can be interpreted as follows:

As $h_1, h_2 \to 0$, the real number:

$$[h_2 ℵ(t, h_1) - h_1 ℵ(t, h_2)]/(h_2 - h_1)$$

approximates $x(t)$ to a higher order than did $\{ℵ(t, h)\}$

Another interpretation.

Set:

$$h_2 = h \text{ and } h_1 = \lambda h$$

$$y(t, h) = [ℵ(t, \lambda h) - \lambda ℵ(t, h)]/(1 - \lambda)$$

Then:

$\{y(t, h)\}$ is a 'better' approximation to $\{x(t)\}$ than $\{ℵ(t, h)\}$

$\{y(t, h)\}$ is the result of transforming $\{ℵ(t, h)\}$ in an algebraic (and group-theoretic/vector bundle-theoretic) way.

Can then be iterated to get a sequence of better-and-better approximations to $\{x(t)\}$. In fact, the whole business has some of the flavor and essence of 'renormalization' and the 'renormalization group' that is so beloved nowadays by Physicists and the Pop-Chaoticians!

GENERALIZED FUNCTION ALGEBRAS AND GEOMETRIC DIFFERENTIAL EQUATION THEORY

NOTES FOR LECTURES

I have spent a good deal of my life thinking about differential equations in a 'geometric way', as far as possible independent of coordinates, analytical machinery, etc. Work on Generalized Function Algebras by Colombeau, Oberguggenberger and Rosinger sets the stage for some fundamental re-thinking of both 'pure' and 'applied' (e.g. mathematial physics, fluid mechanics, control theory) aspects. Important consequences in ways of thinking about and working with nonlinear PDE and ODE! Applications to **Interacting Quantum Field Theory**, **Elementary Particle Physics** and **Fluid Mechanics**. So-called 'Nonlinear Physics'. Also: **Non-smooth control theory** (classical work of Caratheodory and L. C. Young.)

J. F. Colombeau: Multiplication of Distributions: a Tool in Applied Mathematics, Engineering and Physics

M. Oberguggenberger, Multiplication of Distributions and Applications to Partial Differential Equations

E. Rosinger, Nonlinear Partial Differential Equations: An algebraic view of Generalized So;utions.

Colombeau, Heibig and Oberguggenberger, Generalized Solutions to PDE's of Evolution Type, (Preprint).

R. Hermann, Geometric Structures in Nonlinear Physics.

Oberguggenberger-Rosinger Book: Order Completion Methods.

Differential Systems:

> Lie (Founding Father: introduced 'geometric' methods, symmetries, integrability, etc.)

> Cartan (Second Founding Father: Exterior Differential Systems)

> Caratheodory (Control Theory; 'Discontinuous' solutions of differential equations. L. C. Youngs work. Applications to Calculus of Variations, Thermodynamics, etc.).

> Vessiot: Dualized Cartan; 'unified' Cartan and Lie

> Ehresmann: Jetted up and Bundled Cartan and Vessiot.

> Don Spencer: Coming from a background in 'hard' analysis, he started us thinking about general smooth, non-analytic existence theorems for nonlinear PDE motivated by 'geometry'. His 'Princeton-Stranford Programm' to prove non-analytic PDEs have solutions if 'Integrability Conditions' are satisfied.

I first encountered this when Don suggested in 1953 that I 'prove' for my Thesis what later became the Newlander-Nirenberg Theorem about Integrable Almost Complex Structures. Instead, I worked on a very small part of this gigantic whole, involving Lie Groups (a mathematical technology I knew) rather than PDE.

Hans Lewy's Example of smooth (but non-analytic) Linear PDE System without Distribution Solutions (although all 'integrability' conditions are satisfied) showed that 'Don's Programm' was harder than it looked! (It also showed that there were hidden profundities!).

Colombeau has proved that Lewy's Example has solutions in his class of Generalized Functions. Lewy's Example also involves commutation relation of vector fields. Probably there are Fundamental Relations to Control and Lie Theory! Further topic:

> Caratheodory-Chow-Kalman-Hermann-Jurdjevich-Sussman: Controllabity/Accessibillity of systems of ODE's. Extend to Generalized Solutions of ODE' Systems. "Local', 'Infinitesimal' and 'Global' Controllability.

Example of Colombeau-Oberguggenberger-Rosinger Work:

> Set: $D(R^n)$ = differential algebra of smooth (i.e. C^∞) compact-support, real-valued functions on R^n

Embedd $D(R^n)$ in a differential algebra $G(R^n)$ of 'Generalized' Functions on R^n. $G(R^n)$ contains various copies of Schwartz Distributions. Use methods originated by Abraham Robinson in Nonstandard Analysis.

> Fundamental Ideas of:

> **Nature of Infinitely Large and Small!**

Extensions to Manifolds.

> Let X be a Manifold.

> $D(X)$ = algebra of smooth differential forms on X.

> d: $D(X) \rightarrow D(X)$ exterior derivative

> Algebra under exterior multiplcation $\wedge$.

> $U(X)$ = real Lie algebra of smooth vector fields on X.

$\mathbf{D}(X)$ = differential algebra under d and Lie derivative action of $\mathbf{U}(X)$.

$\mathbf{D}(X) \subset \mathbf{G}(X)$ an algebra of Generalized Differential Forms?

Extend Lie derivative action of $\mathbf{U}(X)$, $\wedge$, and d.

An Example of Generalized Function Algebra construction.

Start with $\mathbf{D}(R)$.

$\mathcal{M}(R) = \{$f: f is a map: R x (0, 1] --> R;
 (x, ε) --> $f(x, \varepsilon)$;
 For fixed ε, $\{x \to f(x, \varepsilon)\}$ belongs to $\mathbf{D}(R)$
 For each K compact subset of R,
 $\{\varepsilon \to f(x, \varepsilon)\}$ has at most 'moderate' growth
 in ε as $\varepsilon \to 0$ $\}$

= Functions of Moderate Growth.

Example of Moderate Growth Condition: There exists positive integer m such that:

$|f(x, \varepsilon)\varepsilon^{-m}| < C \varepsilon R$ for $(x, \varepsilon) \varepsilon K \times (0, 1]$, ε suff. small

All derivatives with respect to x also satisfy such a condition.

Simplified version: For each linear differential operator Δ, there exists an integer m > 0 such that $\lim_{\varepsilon \to 0} \varepsilon^{-m} \Delta f = 0$

Construction of GFA as quotient algebra:

Define:

$\mathcal{N}(R) = \{$f: f is a map: R x (0, 1] --> R;
 (x, ε) --> $f(x, \varepsilon)$;

For fixed ε, $\{x \rightarrow f(x, \varepsilon)\}$ belongs to $\mathbf{D}(R)$
For each K compact subset of R,
$\{\varepsilon \rightarrow f(x, \varepsilon)\}$ vanishes to an order greater than 'poynomial' in ε as $\varepsilon \rightarrow 0$.
$\{\varepsilon \rightarrow f(x, \varepsilon)\}$ has 'negligible' growth as $\varepsilon \rightarrow 0\}$

= **negligible functions**

Example of Negligable Growth Condition:

For all positive integers m:

$|f(x, \varepsilon)\varepsilon^{-m}| < C \varepsilon R$ for $(x, \varepsilon) \varepsilon K \times (0, 1]$; ε suff. small
All derivatives with respect to x also satisfy such a condition.

Possible Simplified version: For each linear differential operator Δ, each integer $m > 0$, $\lim_{\varepsilon \rightarrow 0} \varepsilon^{-m} \Delta f = 0$

Set:

$$\mathbf{G}(R) = \mathbf{M}(R)/\mathbf{N}(R)$$

$\mathbf{D}(R)$ is embedded in $\mathbf{M}(R)$ and $\mathbf{G}(R)$:

f: $\{x \rightarrow f(x)\}$ is identified with $\{f(x, \varepsilon) = f(x)\}$

All differential-algebraic operastions extend to $\mathbf{M}(R)$ and $\mathbf{G}(R)$

PDE EXISTENCE THEOREMS IN PAPERS BY COLoMBEAU, HEIBIG AND OBERGUGGENBERGER.
GENERALIZED CAUCHY-KOWALEWSKY THEOREM

$$\partial_t y(x, t) = F(y, y_x, y_{xx}, \ldots)$$

Regularization of PDE used in paper by Colombeau, Heibig and Oberguggenberger, and in Vol. 26 in a Quantum Field Theory context:

Given $\{x \to f(x)\}$.

$$f(x) = \int \delta(x-y)f(y)dy$$

$\{x \to \delta(x)\}$ is Dirac delta function.

$$f_x(x) = \int \delta_x(x-y)f(y)dy$$

Mollify derivatives:. Approximate $\{x \to \delta(x)\}$: $\{x \to \varepsilon^{-1}f(\varepsilon^{-1}x)\}$ where $\{x \to f(x)\}$ is smooth function such that $\int f(x)dx = 1$.

> In a PDE, mollify all derivatives except one: Convert PDE into an Integro-differential equation involving only one derivative. Approximate F (which may be 'generalized' to begin) so that it is bounded, measurable. Use existence theorem for ODE's in Banach space. 'F' on right hand side must be bounded measurable function. Possibility of allowing 'F' to be gen. func. and working with Regularization.

Result: Existence of $\{(x, t) \to y(x, t)\}$ as Generalized Function.

Remark. Quantum Field Theory uses qualitatitvely similar idea to define 'solutions' of Interacting Quantum Field (and Interacting Elementary Particle) Equations!

Example of linear, hyperbolic PDE with gen. fun. coefficients:

Statement of a Theorem (in Oberguggenberger's Book) about linear PDE with Generalized Functions as coefficients:

GFA and DE Systems

Given PDE: $U_t + \Lambda(x, t)U_x = F(x, t)U + G(x, t)U$

$$U|_{\{t=0\}} = A$$

to be solved for $U(x, t)$ as gen. func.

Theorem. Suppose that Λ, F and G are in $\boldsymbol{G}$ (R^2), $A \in \boldsymbol{G}$ (R^1) and, further:

Λ is globally bounded

Λ_x and F are 'locally of logarithmic growth'.

Then, PDE has unique gen. func. solution U.

This result is typical of sort of conditions needed to make machinery work; also, usually extends to non-linear PDE

SPECIALIZATION TO GENERALIZED SOLUTION OF ODE

$$dz/dt = F(z) \; ; \; z \in R^n; \; z(0) = z_0$$
components of F are in $\boldsymbol{G}$ (R).

components of 'solutions': $\{t \longrightarrow z(t)\}$ are in $\boldsymbol{G}$ (R).

First Case:

$$dz/dt = 0.$$

Theorem. Representative of equivalence class of $\{t \longrightarrow z(t)\}$ is:

$$z(t) = z(0) = \{c(\varepsilon) \in R; \text{There exists positive integer m such that: } |c(\varepsilon)\varepsilon^{-m}| < C\}$$

Called **Generalized Constant.**

Proof. Use Mean-Value Theorem to express $\{t \longrightarrow z(t) - z(0)\}$ as integral of derivative z_t. Insert condition that $\{z_t\}$ is negligible.

Existence Theorem for ODE

Choose representatives :

$$\{(z,\varepsilon) \longrightarrow F(z,\varepsilon)\}$$

of Gen. Func. F and initial gen. constant: $\{(t,\varepsilon) \longrightarrow z(0,\varepsilon)\}$. Solve:

$$\partial z/\partial t = F(z, \varepsilon)$$

for $\{(t, \varepsilon) \longrightarrow z(t,\varepsilon)\}$ with:

$$z(t,\varepsilon)|_{t=0} = z(0,\varepsilon).$$

Use Picard ODE Existence Theorem to find $\{(t, \varepsilon) \longrightarrow z(t,\varepsilon)\}$. Differentiability with respect to ε is standard in ODE literature. Work out conditions that $\{(t, \varepsilon) \longrightarrow z(t,\varepsilon)\}$ is 'moderate' as $\varepsilon \longrightarrow 0$. Use Gronwall Lemma. Requires extra condition on F. Existence theorem requires 'locally of logarithmic growth' condition given by Oberguggenberger.

CHOW'S THEOREM/CONTROLLABILITY/FROBENIUS THEOREM IN A GENERALIZED FUNCTION AND MAP CONTEXT.

Review: Suppose given a smooth ODE system:

$$f(\varkappa, d\varkappa/dt) = 0; \varkappa \in R^n$$

$R(\varkappa_0)$ = end-points of 'piece-wise smooth' solutions beginning at $\varkappa_0$

"Reachable Set"

g: $R^n \longrightarrow R$ is **invariant function** if :

$\{t \longrightarrow g(\varkappa(t))\}$ is constant for all solutions $\{t \longrightarrow \varkappa(t)\}$

Generalized Caratheodory- Chow (my version):

$R(z_0)$ = open set iff. there does not exist, locally, a smooth, non-constant invariant function.

Simplest example: n = 3.

$$\theta = dx - A(x)dy - dz$$

$$d\theta \wedge \theta = A_x \, dx \wedge dy \wedge dz$$

$$\aleph(t) = \{ t \rightarrow (x(t), y(t), z(t)) \}$$

ODE System: $\theta \, (d\aleph/dt) = 0$.

'$d\theta \wedge \theta$ not = 0' is condition for reachability.

'Generalized' version:

$$\theta = G(x)dx - A(x)dy - dz$$

G = a Generalized function;

Reachable Set is a set of Generalized Real Numbers.

SMOOTH FLUID-LIKE EQUATIONS ON MANIFOLDS

X = smooth manifold. (Physically: Space-Time)

$U(X)$ = smooth vector fields on X.

D: $U(X) \rightarrow D(X)$ linear differential operator

Fluid Equations for $(V, \omega) \; \varepsilon \; U(X) \times D(X)$:

$$V(\omega) = 0; \; \omega = D(V) \tag{*}$$

$[(V, \omega) \rightarrow V(\omega)$ is Lie derivative operation$]$

$V = \int K(D) * \omega$

Vortices, Shock Waves, etc: (V, ω) solution of (*), but belonging to some Generalized in sense of Colombeau/Rosinger) Extension of $U(X) \times D(X)$.

MODIFICATION OF CANONICAL COMMUTATION RELATIONS IN QUANTUM FIELD THEORY.

Similiarly:

CONVERT PDE'S OF CANONICAL FIELD THEORY INTO INTEGRO-DIFFERENTIAL EQUATIONS AND ODE'S IN BANACH SPACE. MATERIAL IN MY "GEOMETRIC STRUCTURES AND NONLINEAR PHYSICS".

APPLICATIONS TO FLUID/VORTEX FLOW.

THE RELATION BETWEEN THE 'LAGRANGIAN' AND 'EULERIAN' DESCRIPTION OF FLUID FLOW IN THE VORTEX (AND POSSIBLY 'TURBULENT') CASE REQUIRES A THEORY OF GENERALIZED SOLUTIONS OF ODE'S WHOSE RIGHT-HAND-SIDE CONTAINS GENERALIZED FUNCTIONS.

X = manifold with a fixed volume element form

$T(X)$ = tangent vector bundle to X

T = time interval

Eulerian description of fluid flow:

$V: T \times X \rightarrow T(X)$
such that, for $(t, x) \, \varepsilon \, T \times X$,

$V(t, x) \, \varepsilon \, X_x$ = tangent vector space to X at x.

For fixed $t \in T$, let $V(t)$ be the vector field: $\{x \longrightarrow V(t, x)\}$ on X.

'Incompressible Flow' means:

> Lie derivative of the volume element form with respect to each $V(t)$ is zero.

PDE for $\{(t, x) \longrightarrow V(t, x)\}$
> (e.g. Either Euler or Navier-Stokes.)

Non-Smooth Flow (Vortex, Turbulent, ...) means:

> Components of $\{(t, x) \longrightarrow V(t, x)\}$ belong to some Generalized Function class.

Lagrangian orbits:

> Curves $\{t \longrightarrow \aleph(t)\}$ in X such that:
>
> $d\aleph(t)/dt = V(t, \aleph(t))$

MUCH WORK TO BE DONE!

OUTLINE OF RESEARCH IN GENERALIZED FUNCTION ALGEBRA THEORY

1. Introduction.

In Material Science, Fluid Mechanics, and Elementary Particle Physics a prominent role is played by the highly singular solutions of nonlinear PDE's typified by what are classically called 'Vortices' and 'Shock Waves'. In Elementary Particle Physics these topics are modified by the necessity of replacing the underlying PDE's by equations involving quantum mechanical operators. In this Ouline, I will restrict to the 'non-quantum' situation, although it is my intention to go in the 'quantum' direction at a later point. (For example, I believe that the combined analytic and algebraic formalism developed in the books by Colombeau, Oberguggenberger and Rosinger [14-20, 57-61, 62-64] is the first which is well-adapted to capture the Mathematical Mysteries of Renormalization, as in the classic, Nobel-crowned, work of Feynman, Schwinger and Tomonaga.)

This greater flexibility and improved adaptation of the Colombeau-Oberguggenberger-Rosinger Theory to nonlinear situations is due to its very fundamental nature as a **quotient structure**, rather than, as in the Sobolev- Schwartz theory, a linear duality theory. This is akin to the situation in the logical theory of the real numbers, as developed in the Abraham Robinson Theory of Nonstandard Analysis [1, 54, 67]: Indeed, the Colombeau-Rosinger theory is a profound generalization of the Robinson Theory. Both have the property that they require a precise mathematical meaning to 'order of infinity', a mathematical concept which is, I believe, fundamental to the solution of the outstanding problems of Nonlinear Physics (Turbulence, Interacting Elementary Particles, Singularities in Models of Cosmology and Astrophysics, ...).

The Research deals with a broad geometric (and nonlinear!) setting for some of the ideas of Renormalization and Regularization, such as those that lead to such classical concepts as Vortices, Weak Solutions, Shock

Waves, ... In fact, the appropriate algebraic setting has been recently been described (after many tentative attempts going back twenty years!) in a book by Rosinger [64]. Colombeau has pursued a more analytical path [19], and has given a convincing proof of the superiority of his methods over the traditional ones by showing that they apply in non-conservative, plastic-materials situtuations where the traditional methods of 'weak solutions' breaks down. Another important 'applied' insight of the Colombeau-Rosinger theory is that these 'singular' solutions of non-linear PDE's might be **non-unique**. This insight is also, I believe, very important in Elementary Particle Physics, where physicists have only scratched the mathematical surface of the Nonlinear Models they have discovered. The synthesis of these two approaches has been developed in the recent book [57] by Oberguggenberger. I have sketched [45, 46, 47] the extension and application of these ideas to Interacting Quantum Field Theory, Fluid Mechanics, and in fact a broad class of variational problems on Lie pseudogroups. My aim in the next years is:

> a). To extend the setting developed by Colombeau, Oberguggenberger and Rosinger to exterior differential systems [11, 12, 27, 31, 40, 43] and differential systems described in terms of jet bundles.
> b). To pursue the research directions sketched here, particularly the study of Renormalization in the context of the Differential Form Variational Calculus and study the Regularizations and Renormalizations involved in the Feynman 'Quantum Field Theory Integrals' (to distinguish them form the 'Feynman 'Path Integrals') from, on the analytical side, the Colombeau-Rosinger theory and on the geometric side, using ideas of Integral Geometry, of the sort that I have begun in some papers and books written in the 1960's.[35, 36, 38, 47a]
> c). To prepare the way for a NATO Conference on Generalized Function Algebras and Applications, which Michiel Hazcwinkel of the Mathematical Center, Amsterdam and I will direct. I hope that this will be a catalyst for the widespread introduction into the Pure and Applied mathematical and physical worlds of the Generalized Function Algebra ideas and formalism. I travelled to Lyon,

Innsbruck and Amsterdam in March, 1993 to begin this organizational process, and to meet Colombeau, Oberguggenberger and Rosinger. I am planning another such trip in October, 1993 to continue the interaction with them. We certainly plan a comprehensive Proceedings (to be edited by Hazewinkel), for which I plan to write a systematic theory.

2. An 'axiomatization' for the renormalization and generalized function approach to the calculus of variations.

As is appropriate for a subject with its beginnings (with the work of Abraham Robinson) in Logic, I will start with an informal 'axiomatization'.

Let **M** be a set. (The 'model' that I have in mind is that where **M** is a set of smooth mappings between manifolds subject to appropriate boundary conditions.) Let:

$$C(M) \text{ be a set of curves} \qquad (2.1)$$
$$\{\lambda \longrightarrow m(\lambda): \lambda \text{ is a real parameter}\}.$$

Let:

$$A: M \longrightarrow R \qquad (2.2)$$

be a real-valued function on **M**. Physically:

$$A \text{ is the 'action' or 'total energy'.} \qquad (2.3)$$

Definition. The **extremals** are the elements **m** in **M**, such that:

For every curves $\{\lambda \longrightarrow m(\lambda)\}$ in $C(M)$ such that: $\qquad (2.4)$
$$m(0) = m,$$
we have:
$$d[A(m(\lambda))]/d\lambda \big|_{\lambda=0} = 0. \qquad (2.5)$$

I will now define a concept of **generalized extremal**.

Suppose that:

$$M \subset M_{ext}, \quad (2.6)$$

where:

$$M_{ext} \text{ is some class of 'extended maps'}. \quad (2.7)$$

Let:

$$\mathbf{eq} \subset M_{ext} \times M_{ext} \quad (2.8)$$
$$\text{be an equivalence relation on } M_{ext}.$$

Set:

$$M_{gen} = M_{ext}/\mathbf{eq}. \quad (2.9)$$

By composing, we obtain a map:

$$M \dashrightarrow M_{gen} \quad (2.10)$$

which:

$$\text{we suppose is \textbf{one-one}.} \quad (2.11)$$

(This is the **neutrix condition** of Van der Corput and Rosinger [64, 68].)

'**Renormalization**' involves a sort of extension and modification of the Action A. Here is a sketch of how it might appear in this abstract setting.

Postulate that the Action function can be extended to be a function:
$$A_{ext}: M_{ext} \dashrightarrow R_{ext}. \quad (2.12)$$

'R_{ext}' is an 'extension' of the real numbers, e.g. possibly containing 'infinity' in some sense.

However, this function A_{ext} is now 'renormalized' to give a function:

$$A_{ren}: M_{ext} \to R_{ext}. \quad (2.13)$$

We require that:

A_{ren} pass to the quotient to define a map: (2.14)

$$A_{ren}: M_{gen} \longrightarrow R. \qquad (2.15)$$

A_{ren} is now 'extremized' in the usual Calculus of Variations manner to define the set:

$$M_{gen}(\text{extremals}) \subset M_{gen}. \qquad (2.16)$$

The Renormalizations of Quantum Field Theory have a similiar, but more complicated, structure. (Part of the 'complication' is that Quantum Mechanics is involved!) What is to be 'renormalized' is the scattering operator associated with a nonlinear quantum field theory differential operator. In the Schwinger approach, this 'S-Matrix' is also defined by an Action Variational Principle, and there seem to be (at this preliminary stage) similiar tricks of regularizing, and then splitting up the the Action into various parts, some of which are 'discarded' in some complicated recursive way, depending on perturbation theory. The 'self-energy renormalization' of Helmholtz Fluid Vortex Theory (described below) is, perhaps, the simplest example of such a renormalization, and can serve as a Model.

3. The vortex equations as the renormalization of the Euler fluid equations on a Riemannian manifold.

An important and interesting 'model' for this structure is that of **vortices** in incompressable fluid flow. Let X be a Riemannian manifold and let **G** be the Lie algebra of smooth vector fields {V} of divergence zero (i.e. 'infinitesimal volume preserving').

The **energy**:

$E(V)$ of an element of **G** (3.1)

is:

>the integral over X of the square
>(with respect to the inner product on
>tangent vectors defining the metric on X) of V. (3.2)

Let:

>M = a set of smooth curves $\{t \dashrightarrow V(t)\}$ in **G**. (3.3)

The **action**:

>$A(\{t \dashrightarrow V(t)\})$ (3.4)

is:

>the integral over time of $E(V(t))$. (3.5)

The curves $C(M)$ are defined as two-parameter objects (i..e. 'surfaces' in **G**):

>$\{(t,\lambda) \dashrightarrow \{V(t, \lambda)\}$ in **G** (3.6)

satisfying the following **Maurer-Cartan Equations** (or, rather the 'dual' of what are usually defined as these equations):

>$\partial_t(W) - \partial_\lambda(V) = [V, W]$ (3.7)

where:

>$\{(t,\lambda) \dashrightarrow \{W(t, \lambda)\}$ is another 'surface' in **G**.) (3.8)

The **extremals** are then (see [34, 37, 45-47]) the solutions of the **Euler Ideal Incompressible Fluid Equations**:

>$\partial_t(\omega) = L(V)(\omega)$ (3.9)

where: :

>ω the **vorticity**, (3.10)

is:

>a 2-form on X, the result of applying a first order
>linear differential operator to the Eulerian velocity (3.11)

field {V(t)};

$$\omega \longrightarrow L(V)(\omega) \quad (3.12)$$

is the linear operator of Lie derivation of the form ω by the vector field V.

Remark. For the Geometry (and Jet Calculus) of these 'Eulerian Velocity Fields' see my book "Geometry, Physics and Systems" [37].

Of course, these are the equations of 'smooth' flows.

To handle 'turbulence', I propose to look for solutions where {V, ω} lies in some 'generalized function algebra' class, as defined by Colombeau and Rosinger.

The simplest equations of this type seem to be where ω satisfies:

For fixed t, ω is a 'Dirac Delta Function" associated with a submanifold of X. This submanifold varies with t. (3.13)

These objects are **'vortex solutions'**. Of course, since Schwartz distributions cannot be multiplied (and remain within the Schwartz class!) in a natural way, Equation 3.13 makes no sense if 3.9 is satisfied. This is where 'renormalization' comes in!

4. The renormalization of the infinite self-energy of incompressible fluid flows.

To handle the situation where the 'vorticity' ω is outside the class of smooth objects, I follow the ideas of Colombeau and Rosinger cited above and consider 'extended' (but smooth) velocity flows of the form:

$$\{(t,\varepsilon) \longrightarrow V(t,\varepsilon)\} \quad (4.1)$$

The 'ε' are additional parameters. The extended **energy**

$$\varepsilon \longrightarrow E(\{(t,\varepsilon) \longrightarrow V(t,\varepsilon)\}) \tag{4.2}$$

is defined by time-and space integral over the norm of the vector fields $\{V(t,\varepsilon)\}$.

To define 'generalized Eulerian velocity flow' one introduces equivalence relations into the objects of the form 4.1, as explained most systematically in Oberguggenberger's book [58].

Helmholtz Renormalization deals with flows:

$$\{(t,\varepsilon) \longrightarrow V(t,\varepsilon)\} \tag{4.3}$$

which can be written as a sum:

$$V(t,\varepsilon) = V_1(t,\varepsilon) + \ldots + V_n(t,\varepsilon) \tag{4.4}$$
of vector fields in **G** (dependent also on (t, ε)).

The 'energy'/Action:

$$E(\{(t,\varepsilon) \longrightarrow V(t,\varepsilon)\}) \tag{4.5}$$

(as a quadratic form) can be written as a sum:

$$\textbf{self energies + interaction energies} \tag{4.6}$$

The trick is to deal with the flows of the form 4.4 such that:

As $\varepsilon \longrightarrow 0$, **self energy** becomes infinite, (4.7)
whereas **interaction energy** remains finite,
and passes to the quotient to define a real-
valued function on the space of **generalized flows**.

This 'quotient action', physically, **interaction energy**, on the 'generalized flows' can now be extremized in the usual Calculus of Variations way to obtain the **vortex differential equations**.

This formalism can be carried through readily in the 2-D, classical Helmholtz Vortex case, and is more-or-less equivalent (although, in my opinion, much clearer geometrically!) than the traditional renormalization prescription [56] for deriving the Helmholtz story. However, I believe that my method sketched above will be applicable to a much broader class of flows. The payoff will (at least) be an improved understanding of turbulent-like behavior of fluids. Further, once one adopts this Robinson-Rosinger-Colombeau view of the nature of 'infinity', one sees that there is no one **unique** mathematical thing which behaves like 'turbulence', which itself may be an important qualitative insight!

5. Further development of renormalization in the context of the Cartan differential form approach to the variational calculus.

I have published over thirty years [29, 32-34, 37, 38, 40] developments of Cartan's brilliant approach to the Calculus of Variations and associated Mechanics [13]. It can be fit into the Axiomatic framework described in Section in the following way:

$$X \text{ and } Y \text{ are smooth manifolds.} \tag{5.1}$$

I is an exterior differential system on Y, (5.2)
i.e. a differential ideal in the Grassmann algebra of smooth differential forms on Y.

M is a set of smooth maps: $X \dashrightarrow Y$ (5.3)
which are integral submanifolds of I.

The curves $\{m(\lambda)\}$ utilized in Section 2 to define 'extremals' are defined in the following way:

Let:

$$m \; \varepsilon \; M, \tag{5.4}$$
$$\text{i.e. } m^*(I) = 0. \tag{5.5}$$

Let:

$$U(m) = \text{set of smooth vector fields on Y} \tag{5.6}$$
such that:

$$m^*(L(V)(I)) = 0, \tag{5.7}$$

i.e.:

the elements of $U(m)$ are the 'infinitesimal (5.8)
automorphism/deformations' of I at m.

Then:
$$\{m(\lambda)\} \tag{5.9}$$
is defined as:
$$m(\lambda)(x) = \exp(\lambda V)(m(x)). \tag{5.10}$$

The **Action Functional**:
$$A: M \longrightarrow R \tag{5.11}$$
is defined as follows. Let :
$$\theta \text{ be an m-form on Y.} \tag{5.12}$$
$$(m = \text{dimension } X) \tag{5.13}$$

Then,
$$A(m) = \text{integral over } X \text{ of } m^*(\theta) \tag{5.14}$$

If θ is chosen as the so-called **Cartan Form** (as explained in [40, 45]) then it is all the better, since the 'energy' can be defined in terms of θ, as explained in [40].

All of the Variational Principles of Materials Science, e.g. Elasticity, can be described in this framework. I began this description in my books "Differential Geometry and the Calculus of Variations" (which also contains pioneering and so-far unexploited work on Shock Waves) and in "Geometry, Physics, and Systems" (which contains work on Variational Principles of Fluids and Elasticty in the Cartan framework.)

6. Colombeau-Rosinger Theory, the Feynman Quantum Field Theory Integral Renormalizations, Integral Geometry, and 'phisolosphical' remarks.

The Feynman Quantum Field Theory Integral Renormalizations involve computation of the following integral-geometric objects:

$$G = G_1 \cdots G_n \tag{6.1}$$

where:

'G' denotes a matrix of 'Green's Functions' of linear PDE's and: (6.2)

the 'product' in 6.1 is 'matrix product'. (6.3)

Going over to Fourier Transform Land,

$$G \longrightarrow F(G), \tag{6.4}$$

6.1 involves convolutions:

$$F(G) = F(G_1) * \cdots * F(G_n) \tag{6.2}$$

with:

$F(G_1), \ldots, F(G_n)$ = Matrices of Rational Functions on momentum space. (6.3)

$*$ is the convolution product. (6.4)

In addition, there are intricate combinatoric recursion relations between objects of the type 6.2. Indeed, these relations are the mathematical essence of what the physicists call 'renormalizability'. All of this is very complicated, and has already earned many Nobel Prizes. However, there is still important work to be done, and the Colombeau-Rosinger Theory suggests completely new directions and questions. For example:

To what extent are the Generalized Functions obtained by the regularizations chosen independent of the choice? (6.5)

What exactly is the Generalized Function (6.6)

Algebra nature of the objects that the physicists call "Renormalized Green's Functions" and exhibit -via the renormalization algorithm- as some sort of Formal Power Series.

How does one understand the difference between (6.7)
Renormalizable and Unrenormalizable Theories from the Generalized Function Algebra point of view?

Which of the hierarchy of Generalized Function (6.8)
Algebras is the correct one to describe the Interacting Quantum Fields which are at the logical foundation of the Model?

Why must the Sobelev-Schwartz Theory of (6.9)
Distributions be replaced by the Theory of Generalized Function Algebras?

Exactly which of the Generalized Function Algebras (6.10)
is adequate to construct the quantum field theories underlying the Standard Model?

To what extent are the Renormalization Algorithms (6.11)
independent of the algorithms used?

In particular, I want to point out the following question, which might be of extreme importance for the evolution of physics and astronomy, i.e. in the study of 'quantum gravity':

Can the study of at least some 'non-renormalizable (6.12)
quantum field theories'
(e.g. Einsteinian General Relativity)
be interpreted in terms of some of the extended types of Generalized Function Algebras whose theory is now under development by Rosinger and Oberguggenberger [57]?

Turning to 'philosophy' I see many fascinating questions (possibly of great physical importance!) in the non-uniqueness of the Generalized Function Algebra and Nonstandard Analysis models. After all, recall that Robinson's starting point was the axiomatics of the real numbers; the 'nonstandard' examples appeared as Models of these Axioms.

What if the same thing holds for a possible 'Axiomatization' of Interacting Qunatum Field Theory and related Elementary Particles? (6.13)

Geometrically, this might have much to do with a long-conjectured (at least among some who are philosophically minded) 'new' structure of Space-Time 'in the infinitely small'.

Going further:

Why not similiar possibilities for the 'infinitely large' (6.14)

It seems to me that these mathematical possibilities will be needed to explain the mysterious astrophysical and cosmological phenomena being uncovered by advances in observational astronomy.

7. 1-D generalized functions.

In order to provide some mathematical background I will now briefly sketch some simple ideas and situations of GFA theory.

Generalized Function Algebras on the real line:

$$\{x: x \ \varepsilon \ \mathbf{R}\} \tag{7.1}$$

may be considered from the following point of view:

Consider the space **F** of real-valued functions:

$$\{(x, \varepsilon) \dashrightarrow f(x, \varepsilon), x \ \varepsilon \ \mathbf{R}, \varepsilon \ \varepsilon \ \mathbf{R} \text{ and } \varepsilon > 0\} \tag{7.2}$$

such that:

$$\{(x, \varepsilon) \dashrightarrow f(x, \varepsilon)\} \text{ for fixed } \varepsilon \text{ is smooth.} \tag{7.3}$$

Then:

F is an differential algebra under point-wise (7.4)
multiplication and operations by the differential
operators in the variable x.

G, the space of Generalized Functions, will be defined (7.5)
as the quotient of an equivalence relation **EQ**,
i.e.:

$$\mathbf{G} = \mathbf{F} / \mathbf{EQ}. \tag{7.6}$$

EQ will be left unspecified, for the moment, except for the following condition:

EQ is stable under the differential algebra (7.7)
structure on **F**, so that:

G inherits a differential algebra structure. (7.8)

Consider elements:

$$\{(x, \varepsilon) \longrightarrow f(x, \varepsilon)\} \tag{7.9}$$

of **F** generated in the following way:

There is an ε-dependent differential equation in x:

$$D(\varepsilon)(f) = 0 \tag{7.10}$$

such that the element

$$\{(x, \varepsilon) \longrightarrow f(x, \varepsilon)\} \text{ of } \mathbf{F} \tag{7.12}$$

is a solution of "$D(\varepsilon)(f) = 0$".

This point of view gives a traditional, asymptotic-analysis way of looking at the structure of some 1-D generalized function algebras and sheds some light on the traditional quantum-field theoretic renormalization theory.

For example, Consider the 1-D Dirac Delta Function:

$$\{x \longrightarrow \delta(x)\}. \tag{7.13}$$

It may be regared as an element of **G** by assigning to it:

$$\begin{array}{l}\text{the equivalence class of the equivalence relation} \\ \text{(to be specified later) such that is satisfied.}\end{array} \tag{7.14}$$

As the corresponding element:

$$\{(x, \varepsilon) \longrightarrow f(x, \varepsilon)\} \tag{7.15}$$

of **F** choose the following:

$$f(x, \varepsilon) = c\varepsilon^{-1}\exp(-\varepsilon^{-2}x^2). \tag{7.16}$$

The real constant 'c' is chosen so that:

$$c\int_R \exp(-\varepsilon^{-2}x^2)dx = 1. \tag{7.17}$$

In words:

$$\text{"}f(x, \varepsilon) = c\varepsilon^{-1}\exp(-\varepsilon^{-2}x^2)\text{"} \tag{7.18}$$
is the **Gaussian approximation**
to the Dirac function shifted by the
dilation:$\{x \to \varepsilon^{-1}x\}$ on x-space.

Now, differentiate 7.16 with respect to x. (Subscripts denote partial derivatives):

$$f_x = -2c\varepsilon^{-3}x\exp(-\varepsilon^{-2}x^2) = -2\varepsilon^{-2}x\,f \tag{7.19}$$

$$y = f \tag{7.20}$$

is then an ε – dependent of the solution of the differential equation:

$$dy/dx = -2\varepsilon^{-2}x\,y. \tag{7.21}$$

From the point of view of Cartan's Theory of Exterior Differential Systems, we are led to discuss the exterior differential system **E** generated by the one-form:

$$\theta = dy + 2\varepsilon^{-2}x\,y\,dx. \tag{7.22}$$

The map:
$$x \to (x, y(x) = c\varepsilon^{-1}\exp(-\varepsilon^{-2}x^2), \varepsilon) \tag{7.23}$$

is then a **one-dimensional integral manifold** of **E**.

We have:

As $\varepsilon \longrightarrow 0$, **E** goes over to the system $\mathbf{E}_0$ (7.24)
generated by $\theta_0 = x\, ydx$.

The 1-D integral manifolds of the system $\mathbf{E}_0$ are the following curves in $\mathbb{R}^2$, the space of $\{x, y\}$:

$$\{x = 0\} \text{ and } \{y=0, x \neq 0\}. \tag{7.25}$$

Clearly:

> This is a differential-geometric version (7.26)
> of Dirac's Idea that $\{x \longrightarrow \delta(x)\}$ is the
> curve such that $\delta(x) = 0$ if $x \neq 0$,
> but $\delta(0) = \infty$.

> The value '∞' at x=0 is determined by the (7.27)
> normalization condition that the total integral
> is one.

Recall from the work of [47b] that defining:

$$\{x \longrightarrow \delta(x)^2\} \tag{7.28}$$
(and similiar products)

as a Distribution in this way, i.e. via the equivalence class of the map:

$$\{x \longrightarrow [c\varepsilon^{-1}\exp(-\varepsilon^{-2}x^2)]^2\} \tag{7.29}$$

requires a 'subtraction' when integrated against a smooth test-function.

8. Regularization of the ϕ^4 quantum field theory and deformation thery for nonlnear PDE a la Kodaira-Spencer

In my opinion, the elementary particle theory treatise by Cheng and Li [13a] contains the most understandable expostion of quantum field theory renormalization. In Chapter 2, they give what I have found to be the best expostion of the classical theory as developd by Bethe, Feynman, Tomonaga, Dyson and Weinberg. Most important, their formulae make it evident (at least to me) that there is a strong element of Deformation Theory a la Kodaira-Spencer in what they do. I will briefly review some aspects.

Consider the following PDE:

$$D\phi_0 + \mu_0^2 \phi_0 + \lambda_0^2 \phi_0^3 = 0. \tag{8.1}$$

$$\text{D is a linear, second-order differential operator.} \tag{8.2}$$

The 'solution':

$$x \longrightarrow \phi_0(x, \mu_0, \lambda_0, \varepsilon) \tag{8.3}$$
$$\textbf{is the 'bare' quantum field;}$$

$$\mu_0, \lambda_0 \text{ the 'bare' coupling constants,} \tag{8.4}$$

$$\varepsilon \text{ some sort of cut-off parameter.} \tag{8.5}$$

The 'renormalized field' is to be defined a la Rosinger-Colombeau-Oberguggenberger as the quotient of such things under appropriate equivalence relations. Of course, finding these equivalence relations will be the main problem of the theory. Looking at the Cheng-Li formulas gives some clues.

Geometrically, this involves a vector bundle and action of various dilation groups acting. For example, the 'renormalization constant':

Z acts via:

$$\phi_0(x, \mu_0, \lambda_0, \varepsilon) \longrightarrow Z^{-1/2} \phi_0(x, \mu_0, \lambda_0, \varepsilon). \tag{8.6}$$

There are also the dilations acting on the mass and coupling constant parameters. The asymptotic limit (which is what physicists call 'renormalization') invoves going to the boundary in the compactification of an appropriately defined variety in the $(Z, \mu_0, \lambda_0, \varepsilon)$-space by various Feynman diagrams. (Although my aim will be to do this more intrinsically, without using Feynman diagrams. This variety also depends on another discrete parameter, the order of the diagram.) Underlying all this is, no doubt some sort of Cohomology associated with the appropriate Deformation Theory.

Although this Model is essential for understanding what the physicists are doing, I prefer to recast in terms of Deformation of an 'algebraic structure'. What precisely is that 'algebraic structure' remains to be discovered.

9. A closing statement.

I have been privileged to witness the introduction of many new and exciting ideas in mathematics, physics and engineering. One such period was my graduate student years (in 1952-55) in Princeton, Paris, Strasbourg and Amsterdam. New methods of differential geometry and Lie group theory were introduced; sheaf and fiber bundle theory and associated homology and homotopy theory were developed; and my thesis advisor Don Spencer began his work recasting the analytical and algebraic theory of differential systems. A second such period was my brief (1959-61) experience as a Staff Mathematician at Lincoln Laboratory of MIT. In contrast to the first period, this was an exciting time for the development of Applied Mathematics. This was the beginning of the Space Program, and a new generation of engineers were beginning to tackle problems that not yet been modernized, emphasizing the use of computers, methods of signal processing, control and filter theory, etc. In addition, there was a new and improved collection of mathematical tools -probability and stochastic theory, linear algebra, dynamical systems, etc. - available to think about the engineering problems in more mathematical terms. I was on the scene to suggest application of differential-geometric methods to handle the sort of nonlinear problems that were on the horizon. My third experience came in 1962-69 with the application of Lie group theory and differential geometry to elementary particle physics.

I believe that we are on the brink of a new burst of such a creative research breakthough; this time we may gain fundamental new insights into **nonlinear physics** using the new Colombeau-Rosinger-Oberguggenberger insights.

Bibliography

1. S. Albeverio, J. Fenstad, R. Hoegh-Krohn and T. Lindstrom, **Nonstandard Methods in Stochastic Analysis and Mathematical Physics,** Academic Press, 1986.

2. J. Aragona and H. A. Biagioni, An intrinsic definition of the Colombeau algebra of generalized functions, to appear, *Analysis Mathematica.*

3. H. A. Biagioni, **A Nonlinear Theory of Generalized Functions,** Springer-Verlag, 1990

4. H. A. Biagioni, The Cauchy problem for semilinear hyperbolic systems with generalized functions as initial conditions, *Results in Mathematics*, 14, 1988, 231-241.

5. H. A. Biagioni and J. F. Colombeau, Borel's theorem for generalized functions, *Studia Mathematica,* 1985, 179-183

6. H. A. Biagioni and J. F. Colombeau, New generalized functions and C^∞ functions with values in generalized complex numbers, *J. London Math. Soc.,*33 (1986), 169-179.

7. H. A. Biagioni and J. F. Colombeau, Whitney's extension theorem for generalized functions, *Mathematical Analysis and Applications,* 114 (1986), 574-583

8. H. A. Biagioni and M. Oberguggenberger, Generalized solutions to Burger's equation, *Journal of Differential Equations*, 97 (1992), 263-287.

9. H. A. Biagioni, On Treves' result of approximate solutions to a nonlinear first-order system, preprint.

10. H. A. Biagioni and M. Oberguggenberger, Generalized solutions to the Korteweg-deVries and the Regularized Long-Wave Equations, *SIAM J. Math. Appl.*, 23 (1992), 923-940

11. R. Bryant, S. Chern, R. Gardner, H, Goldschmidt and P. Griffiths, **Exterior Differential Systems**, Springer-Verlag, New York, 1991

12. E. Cartan, **Les systemes differentielles exterieures et leurs applications geometriques**, Hermann, Paris, 1946.

13. E. Cartan, **Lecons sur les Invariants Integraux**, Hermann, Paris, 1922.

13a). Ta-Pei Cheng and Ling-Fong Li, **Gauge Theory of Elementary Particles,** Oxford Univ. press, 1984.

13b). A. Chorin and J. Marsden, **A Mathematical Introduction to Fluid Mechanics.**, Springer-Verlag, 1990

14. J. F. Colombeau, Multiplication of distributions, *Bull. Am. Math. Soc.*, 23, 25-268, 1990.

15. J. F. Colombeau, **New Generalized Functions and Multiplication of Distributions,** North-Holland, Amsterdam, 1984

16. J. F. Colombeau, **Elementary Introduction to New Generalized Functions,** North-Holland, Amsterdam, 1985

17. J. F. Colombeau and A. Y. LeRoux, Multiplication of distributions in elasticity and hydrodynamics, *J. Math. Physics*, 29, 315-319, 1988.

18. J. F. Colombeau, The elastoplastic shock problem as an example of the resolution of ambiguities in the multiplication of distributions, *J. Math. Physics*, 30, 2273-2279, 1989

19. J. F. Colombeau, **Multiplication of Distributions: a Tool in Applied Mathematics, Engineering and Physics,** Springer-Verlag, 1992.

20. J. F. Colombeau, A. Heibig and M. Oberguggenberger, Generalized solutions to partial differential equations of evolution type, preprint, *Ecole Normale Superieure de Lyon*, 1991.

21. N. Cutland, Ed., **Nonstandard Analysis and its Applications,** Cambridge University Press, 1988.

22. J. W. de Roever and M. Damsma, Colombeau algebras on a C^∞ manifold, *Indag. Mathem., N.. S.*, 2, (1991), 341-358

23. R. Di Perna, Measure Valued solutions to consevation laws, Arch. Rat. Mech. Anal. **88**, 1985, 223-270

24. R. Di Perna and A, Majda, Oscillations and concentrations in weak solutions of the incompressible fluid equations, Comm. Math. Physics, **108**, 1987, 667-689

25. Yu. V. Egorov, A contribution to the theory of generalized functions, *Russian Math Surv.*, 45 (1990), 1-49

26. H. G. Embacher, G. Grubl, and M. Oberguggenberger, products of distributions in several variables and application to zero-mass QED_2, Zeitschrift fur Analysis und ihre Anwedungen, 11 (1992), 437-454.

26a). M. Goze, Perturbations of Lie Algebra Structures, in: M Hazewinkel and M. Gerstenhaber, Eds., *Deformation Theory of Algebras and Structures and Applications,* Kluwer, 1988, pp. 265- 355

27. P. A. Griffiths, **Exterior Differential Systems and the Calculus of Variations,** Birkhauser, Boston, 1983.

28. P. A. Griffiths and G. Jensen, **Differential Systems and Isometric Embeddings**, Princeton Univ. Press, 1987.

29. R. Hermann, Some Differential Geometric Aspects of the Lagrange Variational Problem, *Illinois J. Math.* 6 (1962), pp. 634-673.

30. R. Hermann, On the Accessibiltiy Problem of Control Theory, *Proc. of the Symposium on Differential Equations*, J. Lasalle and S. Lefschetz, Eds., Colorado Springs, Academic Press, 1961.

31. R. Hermann,. E. Cartan's Geometric Theory of Partial Differential Equations, *Advances in Math* 1 (1965), pp. 265-317.

32. R. Hermann, The Second Variation for Variational Problems in Canonical Form, *Bull. Amer. Math. Soc.* Vol 71 (1965), pp. 145-148.

33. R. Hermann, The Second Variation for Minimal Submanifolds, *J. of Math. and Mech.* 16 (1966), pp. 473-492.

34. R. Hermann, **Differential Geometry and the Calculus of Variations,** Academic Press New York 1969. Second Edition, Math Sci Press, 1977.

35. R. Hermann, **Lie Algebras and Quantum Mechanics,** W. A. Benjamin New York 1971. 400 pp.

36. R. Hermann, **Vector Bundles in Mathematical Physics,** Parts I and II, W. A. Benjamin, New York 1970. 441 pp. and 400 pp.

37. R. Hermann, **Geometry Physics and Systems,** Marcel Dekker New York 1973.

38. R. Hermann, *Bohr-Sommerfeld Quantization in General Relativity and other Nonlinear Field and Particle Theories,* Proceedings of the 1979 Conference on Quantum Gravity, Loyola Univ., New Orleans, R. A. Marlow, ed. Academic Press.

39. R. Hermann, Geometric Theory of Deformation and Linearization of Pfaffian Systems and Its Application to System Theory and Mathematical Physics, *J. Math Phys.* 24, 2268-2276, 1983.

40. R. Hermann, Differential Form Methods in the Theory of Variational Systems and Lagrangian Field Theories, *Acta App. Math.*, 12, 1988, 35-78.

41. R. Hermann, Geometric Construction and Properties of Some families of Solutions of Nonlinear Partial Differential Equations, *J. Math. Phys.*, 24, 1983, 510-521.

42. R. Hermann,. Perturbation and Linearization of Nonlinear Control Systems, in: R. Hunt and C. Martin(eds.), *Proc. Berkeley-Ames Conf. on Nonlinear Problems in Control and Fluid Mechanics*, Math Sci 43.

43. R. Hermann, Geometric and Lie-Theoretic Principles in Pure and Applied Deformation Theory, in: M. Gerstenhaber and M. Hazewinkel (eds.), *Deformations of Algebras and Applications*, D. Reidel, Dordrecht, 1988, 701-796.

44. R. Hermann, Perturbation Theory for Nonlinear Feedback Control Systems and Goldschmidt-Spencer Integrability of Linear Partial Differential Equations, *Acta App. Math.*, 18, 17-57, 1990

45. R. Hermann, **Geometric Structures and Nonlinear Physics,** Math Sci Press, 1992

46. R. Hermann, Generalized Function Algebras and Nonlinear Physics, to appear, *Proceedings of the III International Conference on Stochastic Processes, Physics and Geometry*, S. Albeverio, Ed.

47. R. Hermann, **Constrained Mechanics and Lie Theory**, Math Sci Press, 1993

47a). R. Hermann, Analytic Continuation of Group Reresentations, Parts I-VI, *Comm. Math. Physics*, 2, 251-270, 1966; 3, 53-74, 1966; 4, 75-97, 1966; 5, 131-156, 1967; 5, 157-190, 1967; 6, 205-225, 1967.

47b). R. Hermann, Geometric Ideas in Lie Group Harmonic Analysis Theory, *Proc. of the Washington U. Symposium on Symmetric Spaces*. Eds:, W. Boothby and G. Weiss, Marcel Dekker Inc. New York. 1972.

47c. R. Hermann, A Geometric Formula for Current Algebra Commutation Relations, *Phys. Rev.* 177 (1969), p. 2449.

47d. R. Hermann, *Quantum Field Theories with Degenerate Lagrangians*, Phys. Rev. 177 (1969), p. 2453.

47e. R. Hermann, *Algebraic and Geometric Structures in Current Algebra Theory* ONR Technical Report No. 1 Institute for Advanced Study Princeton, New Jersey 1969

47f. R. Hermann, Current Algebra, Sugawara Model and Differential Geometry, *J. of Math. Phys.* 11 (1970), pp. 1825-1829.

47g. R. Hermann, *Infinite Dimensional Lie Algebra and Current Algebra*, Proc. of the 1960 Battelle-Seattle Recontres on Math Physics, Lecture Notes in Physics, Springer-Verlag,1970 pp. 212-237.

47h. R. Hermann, *Physical Aspects of Lie Group Theory.*, University of Montreal Press, Montreal 1974.

47i. R. Hermann, *Topics in the Mathematics of Quantum Mechanics,* Vol. VI of Interdisciplinary Mathematics Math Sci Press Brookline, Mass. 1973.

47j. R. Hermann, Compactifications of Homogeneous Spaces and Contractions of Lie Groups, **Proc. Nat. Acad. Sci.** 51 (1964), pp. 456-461.

47k. R. Hermann, Complex Domains and Homogeneous Spaces, **J. Math. and Mech.** 13 (1964), pp. 667-672.

471. R. Hermann, A Geometric Formula for Current Algebra Commutation Relations, **Phys. Rev.** 177 (1969), p. 2449.

48. C. Itzykson and J. Zuber, **Quantum Field Theory**, McGraw-Hill, 1980

48a. J. Jauch and M. Reed, Nonlinear superposition and aborbtion of delta waves in one space dimension, J. Func. Anal. 73 (1987), 152-178

49. K. Kodaira, **Complex Manifolds and Deformations of Complex Structures**, Springer-Verlag, 1986.

50. K. Kodaira and D. C. Spencer, On deformations of complex analytic structures, *Ann. Math.* 67, 328-466, 1958.

51. A. Kumpera and D. C. Spencer, **Lie Equations**, Ann. of Math. Studies 73 (Princeton U. P., Princeton, NJ, 1972).

52. W.Lamb, Hydrodynamics, Cambridge Univ. Press, 1932

53. A. Y. Le Roux and P. De Luca, A velocity-pressure model for elastodynamics, in "Nonlinear Hyperbolic Equations, Theory, Compuational Methods and Aplications", J. Ballman and R. Jetch, Editors, Notes on Numerical Fluid Mechanics, vol. 24, Vieweg Verlag 1989, pp. 374-383.

54. A. H. Lightstone and A. Robinson, **Nonarchimedean Fields and Asymptotic Expansions**, North-Holland, 1975.

55. J.J. Lodder, **Towards a symmetrical theory of generalized functions**, CWI Tract # 79, Amsterdam. 1991.

56. J. Marsden and A. Weinstein, Coadjoint Orbits, Vortices and Clebsch Variables for Incompresible fluids, *Physica* 7D, 305-323, 1983.

57. M. Oberguggenberger, **Multiplication of Distributions and Applications to Partial Differential Equations**, Pitman, 1992.

58. M. Oberguggenberger and E. Rosinger, **Solution of Continuous Nonlinear PDE's through Order Completion**, Technical Report, University of Pretoria, 1991.

59. M. Oberguggenberger and Ya-Guang Wang, Generalized solutions to conservation laws, preprint.

60. M. Oberguggenberger, Case study of a nonlinear, nonconservative, nonstrictly hyperbolic system, *Nonlinear analysis, Theory, Methods and Applications*, 19 (1992), 53-79.

61. M. Oberguggenberger, Nonlinear theories of generalized functions, preprint.

62. E. E. Rosinger, **Generalized Solutions to Nonlinear PDE**, North-Holland, Amsterdam, 1987

63. E. E. Rosinger, **Nonlinear Partial Differential Equations**, North-Holland, Amsterdam, 1980

64. E. E. Rosinger, **Nonlinear Partial Differential Equations, an algebraic view of generalized solutions**, North-Holland, Amsterdam, 1990

65. D. C. Spencer, Overdetermined systems of linear partial differential equations, *Bull. Am. Math. Soc.* 75, 179-239 (1969).

66. D. C. Spencer, Deformation of structures on manifolds defined by transitive, continuous pseudogroups, *Ann. Math.* 76, 306-445, 1962

67. K. D. Stroyan and W. A. J. Luxemburg, **Introduction to the Theory of Infinitesimals**, Academic Press, 1976.

68. J. G. Van der Corput, Introduction to Neutrix Calculus, *J. d'Analyse Math.*, 7 (1959), 281-398.

69. K. Yang, **Exterior Differential Systems and Equivalence Problems,** Kluwer, 1992

CHAPTER 1

THE LIE-THEORETIC NUMERICAL ANALYSIS OF SOME FROBENIUS INTEGRABLE DIFFERENTIAL SYSTEMS

1. Introduction.

In my book, "Geometric Computing Science: First Steps" [3], I attempted to provide a mathematical overview and description of computer programs from a 'geometric' point of view. In the short term, the goal of covering a major part of this area has proved to be overly ambitious. My research in the past few years in the direction of Computing Science has focussed on a small piece of this whole, namely understanding the geometric nature of differential system numerical analysis algorithms. This Chapter is a preliminary report on this Research, supported in part by an Exploratory Research grant from the Computational Mathematics Program of NSF. I have also benefited from the hospitality of the AI Lab of MIT and conversations with Prof. Gerald Sussman.

My plan for this Chapter is to assume known some of the basic analytical facts [1] about ordinary differential equation numerical analysis, and to develop additional directions of research motivated by the contemporary theory of Differential Systems and Deformation Theory [2]. In order to keep on as concrete a level as possible, I will emphasize the theory of Frobenius Integrable PDE systems in two independent variables. I will begin with certain aspects of Numerical Analysis of linear ordinary differential equations with constant coefficients, developing what I will call a WKB-Type of representation of the numerical approximations. I will also briefly mention in this paper the classical technique of constructing numerical approximations to dynamical systems via the **Richardson Method of Extrapolation or Deferred Limit** [1, 5] and relations to what the physicists call **Lattice Gauge Theory** [6].

In the mathematical background of my work is the idea that the process of numerical approximation of solutions of differential systems - and much else in the computer science descripion of the continuous 'world' in terms of 'discrete' mathematical structures - uses mathematical processes which are akin to those encounted in Mathematical Physics and Differential Geometry, involving the 'deformation' of one geometric or algebraic structure into another. The main deformation structures involved are the deformations of groups and their homomomorphisms. I hope in a later work to go more deepy into the relation between differential system numerical analysis and the deformation theory of algebraic and geometric structures.

In this paper, I will deal with two types of differential systems: The first is an ODE of the form:

$$dx/dt = f(x); \quad x \in \mathbf{R}^n; \tag{1.1}$$
$$f \text{ is a smooth map: } \mathbf{R}^n \longrightarrow \mathbf{R}^n$$

I also consider Frobenius Integrable system of two PDE's in two real variables. I will now describe a simple example of such a system. Suppose given the following data:

$$x \in \mathbf{R}^n; \quad s, t \in \mathbf{R} \tag{1.2}$$

$$\text{Smooth maps } f, g: \mathbf{R}^n \times \mathbf{R}^2 \longrightarrow \mathbf{R}^n, \tag{1.3}$$

Associate with 1.2-1.3 the following PDE system:

$$x_t = f(x, s, t) \tag{1.4}$$

$$x_s = g(x, s, t). \tag{1.5}$$

1.4-1.5 is to be solved for a map: $(s, t) \longrightarrow x(s, t)$ of $\mathbf{R}^2 \longrightarrow \mathbf{R}^n$. (Subscripts denote partial derivatives.)

To define the concept of 'Frobenius Integrability' of the PDE system 1.4-1.5, associate with it the following vector fields on $R^n \times R^2$:

$$V = f(x, s, t)\partial/\partial x + \partial/\partial t \qquad (1.6)$$

$$W = g(x, s, t)\partial/\partial x + \partial/\partial s \qquad (1.7)$$

Definition. The PDE System 1.4-1.5 is **Frobenius Integrable** if the following relation is satisfied:

$$[V, W] = 0, \qquad (1.8)$$

where [,] denotes the Jacobi-Lie Bracket of vector fields on $R^n \times R^2$

Remark. The PDE system 1.4-1.5 can also be interpreted geometrically as a **connection in a fiber bundle** whose base is R^2. Thus the Frobenius Integrability Conditions 1.8 can be interpreted geometrically as the **vanishing of the curvature of this connection**. Although I will not emphasize this point of view in this paper, it is worth noting, if only to establish a relation between what I am doing here and what the physicists call **Lattice Gauge Theory** [6].

The PDE systems of type 1.4-1.5, satisfying 1.8, belong to that class of PDE systems which can be completely solved using algorithms involving only ordinary differential equations. There was extensive work in the late 19th and early 20th century by such mathematicians as Cartan, Goursat and Lie delineating this class of PDE's; much of this work has turned out to be relevant to the contempory theory of Integrable Differential Systems. It seems to me that it would be a scientifically worth-while program to develop the Theory and Practice of Algorithms for describing the solutions of such differential systems in terms of contempory computing science. (For example, such methodology would be useful in describing rotatating bodies and other kinematical and dynamical situations which involve Lie groups.) This paper may be considered as an initial step towards this goal.

In this paper, I will emphasize the case where the differential systems 1.1 and 1.4-1.5 consist of linear differential equations. However, at the end of it I will sketch some ideas for extending the methods, using Lie theory, to the nonlinear situations.

2. Linear time-invariant ODE's and the Euler approximation.

Consider first a linear ODE of the form:

$$dx/dt = Ax, \tag{2.1}$$

$$x \in R^n \tag{2.2}$$

$$A: R^n \rightarrow R^n \text{ is a linear map,} \tag{2.3}$$

or:

$$A \text{ is an } n \times n \text{ real matrix.} \tag{2.4}$$

The solution of 2.1 can of course be written in the following form:

$$x(t) = \exp(tA)(x(0)), \tag{2.5}$$

with:

$$\exp(tA) = \Sigma_j t^j A^j / j! \tag{2.6}$$

The Euler Formula is:

$$\exp(tA) = \lim_{n \to \infty} (1 + At/n)^n \tag{2.7}$$

Set:

$$h = t/n \tag{2.8}$$

In the Numerical Analysis context, h is the **step size**. For fixed t, we then have:

$$x(t) = \lim_{h \to 0} (1 + Ah)^{t/h}(x(0)) \tag{2.9}$$

For each m, we seek an expansion of 2.9 the following form:

$$x(t) = \sum_{0 \leq j \leq m} x_j(t) h^j + R_m(t, h) h^{m+1} \qquad (2.10)$$

In general, I will call asymptotic expansions of type 2.10 the **Richardson Expansion** associated with the ODE 2.1.

Remark. I use this name because expansions of the type 2.10 are the input to what is called **Richardson Extrapolation Method** in the Numerical Analysis literature. Richardson's work [5] - and its modern extensions [1] - involves algebraic operations on series of the type 2.10 to give new expansions of the same type.

My first goal is to derive - by simple Calculus methods - explicit expansions of the form 2.10 for this linear case for the Euler formula 2.9.

3. Expansions of the Euler approximation of linear ODE's based on a 'WKB' type of integral representation of the solutions of linear, time-invariant difference equations.

The basic idea of the Euler Method for Numerical Analysis of the ODE 2.1 is to approximate the one-parameter group $\{t \to \exp(tA)\}$ of n x n matrices by the family of one-(discrete) parameter semigroups $\{n \to (1+ hA)^n\}$. 'h' is an additional parameter, usually identified with the **step size**. In order to exhibit this approximation in a convenient analytic form, I will use the following elementary calculus formula:

$$\log([1+a]^b) = \int_{[0,1]} ab[1+au]^{-1} du \qquad (3.1)$$

which holds for real 'a' and 'b'.

Let A again be a real, nxn matrix and 'h' a real number. Substitute: $\{a \to Ah\}$ (for h sufficiently small) and $\{b \to n\}$ in 3.1, obtaining:

$$\log([1+Ah]^n) = \int_{[0,1]} Ahn[1+Ahu]^{-1} du \qquad (3.2)$$

and:
$$[1+Ah]^n = \exp(Ahn\int_{[0,1]} [1+Ahu]^{-1} du) \qquad (3.3)$$

Theorem 3.1. For each real t,
$$\lim_{h \to 0} [1+Ah]^{t/h} = \exp(At) \qquad (3.4)$$

3.4 is equivalent to the Euler Formula 2.7.

Proof. Substitute:
$$n = t/h \qquad (3.5)$$

into 3.3, obtaining:
$$[1+Ah]^{t/h} = \exp(At\int_{[0,1]} [1+Ahu]^{-1} du) \qquad (3.6)$$

3.4 now follows by the Lebesgue Dominated Convergence Theorem, applied, as h --> 0, to the right hand side of 3.6.

q.e.d.

Theorem 3.2. The Euler Formula can be written in the following form:
$$\exp(tA) = \lim_{h \to 0} \exp t\left[\int_{[0,1]} [A/(1+Ahu)] du\right] \qquad (3.7)$$

Proof. Combine 2.7 and 3.6.

Remark. Formula 3.7 has an important deformation-theoretic meaning. Namely, it says that, in some appropriate sense, the homorphism: {t --> exp(tA)} of the Lie group R of additive real numbers is 'deformed' (depending on the parameter 'h') by the one- parameter family:
$$\{t \to \exp(t\left[\int_{[0,1]} [A/(1+Ahu)] du\right])\} \qquad (3.8)$$

of group homomorphisms. This sort of deformation is the analog of the deformation-of-Lie-algebra-homomorphism theory constructed by Nijenhuis and R. Richardson [5]

Theorem 3.3. The matrix-valued function $\{h \rightarrow [1+Ah]^{t/h}\}$ is analytic in the neighborhood of 'h=0'. Here is an expansion about 'h=0':

$$[1+Ah]^{t/h} = \exp\left[t A \Sigma_j (-1)^j A^j h^j / (j+1)\right] \quad (3.6)$$

Proof. For fixed u, the matrix-valued function $[h \rightarrow A/(1+Ahu)]$ is analytic in the neighborhood of 'h=0', hence the integral with respect to u over [0, 1] is also analytic. Write the power-series expansion of the integrand of 3.6 in the following form:

$$1/(1+Ahu) = \Sigma_j (-1)^j A^j u^j h^j \quad (3.7)$$

Integrating term-by-term with respect to u over the interval [0, 1] proves 3.6.

q.e.d.

We can now use Theorem 3.3 to prove the analyticity about 'h=0' of the Euler expansion of the linear ODE 1.1.

Theorem 3.4. The matrix-valued map:

$$h \rightarrow (1+Ah)^{1/h} \quad (3.8)$$

is real-analytic about $\{h = 0\}$ in the sense that the well-defined values at:

$$h = 1, 1/2, 1/3, \ldots \quad (3.9)$$

can be extended to a map in a neighborhood of 'h= 0' in such a way as to be real analytic. We have:

$$(1 + Ah)^{1/h} = \exp\left[\Sigma_j(-1)^j A^{j+1} h^j/(j+1)\right] \qquad (3.10)$$

Proof. Follows from 3.6.

q.e.d.

Remarks. Notice the resemblence between formulas 3.10 and 3.6 and the WKB Formula of Mathematical Physics. 3.10 also tells us that the Euler Approximation is a real-analytic function in the step-size parameter 'h'. The general existence theorem for ODE numerical analysis (see Gear's book [1], Theorem 4.3) only provides, in general, an asymptotic expansion in 'h'. It would be very interesting to delineate further which numerical algorithms lead to such an analyticity property.

Theorem 3.5. Alternately, we can write the first few terms of the power series expansion of $\{h \rightarrow (1 + Ah)^{1/h}\}$ in the following form:

$$(1 + Ah)^{1/h} = \exp(A)\left[1 - A^2 h/2 + \ldots\right] \qquad (3.11)$$

Proof. Let us rewrite 3.1 and 3.5 as follows:

$$(1 + Ah)^{1/h} = \exp\left[\int_{[0,1]} [A/(1+ Ahu)]du\right] \qquad (3.12)$$

Then:

$$d\left[(1 + Ah)^{1/h}\right]/dh =$$

$$-\exp\left[\int_{[0,1]} [A/(1+ Ahu)]du\right]\left[\int_{[0,1]} Au/(1+ Ahu)^2 du\right] \qquad (3.13)$$

$$d\left[(1 + Ah)^{1/h}\right]/dh\bigg|_{h=0} =$$

$$-\exp A\left[\int_{[0,1]} Au\, du\right] = -A^2/2 \exp A \qquad (3.14)$$

q.e.d.

Let us now examine more general linear difference equation approximations and deformations to linear differential equations of the type 1.1.

4. More general approximations of the WKB type for linear, time-independent ODE's.

Let us now consider a linear difference equation of the following form:

$$\aleph_{app}([n+1]h) = (1 + hB(h))(\aleph_{app}(nh)) \qquad (4.1)$$

where:

$\{h \longrightarrow B(h)\}$ is a smooth map of an interval $(-a, a)$ (4.2)
of real numbers into the Lie algebra of $n \times n$
real matrices.

$$B(0) = A \text{ for all } h. \qquad (4.3)$$

Definition. The h-dependent family of difference equations 4.1 is said to be a **deformation** of the differential equation 2.1 if 4.2-4.3 is satisfied.

We can of course readily solve the linear difference equation 4.1 by iteration:

$$\aleph_{app}(nh) = (1 + hB(h))^n(\aleph(0)) \qquad (4.4)$$

As in Section 3, we can derive a WKB-type representation of the semigroup $\{n \longrightarrow (1 + hB(h))^n\}$ providing the following condition is satisfied:

$$[B(h), B(h')] = 0 \text{ for all } h, h' \ \varepsilon \ (-a, a) \qquad (4.5)$$

The assumption 4.5 now enables us to substitute: {a --> hB(h)} in 3.1 to obtain the following formula:

$$\log((1 + hB(h))^n = n\int_{[0, 1]} [1 + uhB(uh)]^{-1} [hB(h)] du \qquad (4.6)$$

Take the exponential of the nxn matrix on both sides of 4.6:

$$(1 + hB(h))^n = \exp(n\int_{[0, 1]} [1 + uhB(uh)]^{-1} [hB(h)] du) \qquad (4.7)$$

Introducing the variable 't' as:

$$t = hn, \qquad (4.8)$$

leads to the following:

Theorem 4.1. With 4.5 and 4.8 satisfied, the semigroup {n --> $(1 + hB(h))^n$} admits the following WKB-type representation:

$$(1 + hB(h))^n = \exp(t\int_{[0, 1]} [1 + uhB(uh)]^{-1} [B(uh))] du \qquad (4.9)$$

Remark. Further asymptotic expansions of {t --> exp(tA)} in powers of h may be obtained by substituting into 4.9 asymptotic expansions of {h --> B(h)}

I will now work through this analysis more explicitly for the case arising from the 'centered difference approximation' of the linear ODE 2.1.

5. The centered-difference approximation for linear ODE's and the Cayley transform of the A-matrix.

Consider the difference-equation approximation of 2.1 obtained by using a centered- difference approximation for the unknown function {t --> א(t)} and its derivatives:

$$א(t) \approx [א(t+h/2) + א(t-h/2)]/2 \qquad (5.1)$$

Lie-Theoretic Numerical

$$[d\varkappa/dt](t) \approx [\varkappa(t+h/2) - \varkappa(t-h/2)]/h \tag{5.2}$$

If $\{t \rightarrow \varkappa(t)\}$ is a solution of 2.1, substituting 5.1 and 5.2 into 2.1 leads to the following difference equation:

$$[\varkappa_{app}(t+h/2) - \varkappa_{app}(t-h/2)]/h = A[\varkappa_{app}(t+h/2) + \varkappa_{app}(t-h/2)]/2 \tag{5.3}$$

which can also be written in the following form:

$$\varkappa_{app}(t+h/2) = [1 - Ah/2]^{-1}[1 + Ah/2]\varkappa_{app}(t-h/2) \tag{5.4}$$

This heuristic leads us to examine the matrix-valued function:

$$[1 - Ah/2]^{-n}[1 + Ah/2]^n \tag{5.5}$$

and compare it asymptotically, for:

$$t = nh; \; h \rightarrow 0 \tag{5.6}$$

with: $\{t \rightarrow \exp(tA)\}$.

Recall the following:

Definition. Given the nxn matrix A, the matrix:

$$[1 - A]^{-1}[1 + A] \tag{5.7}$$

is called the **Cayley Transform** of A.

Theorem 5.1. Set:

$$B(h) = A[1 - Ah/2]^{-1} \tag{5.9}$$

Then:

$$[1 - Ah/2]^{-1}[1 + Ah/2] = 1 + hB(h) \tag{5.10}$$

Proof. Substitute 5.9 into 5.10 and verify that the resulting relation is an identity.

q.e.d.

Theorem 5.2. With $t = nh$, the following WKB representation of 5.5 holds:

$$[1 - Ah/2]^{-n}[1 + Ah/2]^n = \exp(At\int_{[0,\,1]}[1 - uhA/2][1 - A^2 u^2 h^2/4]^{-1} du) \tag{5.11}$$

Proof. Using 5.9, we have:

$$1 + hB(h) = 1 + hA/(1-Ah/2) = [(1-Ah/2) + hA](1-Ah/2)^{-1}$$

$$= [(1 + hA/2](1-Ah/2)^{-1}$$

hence:

$$[1 + hB(h)]^{-1} = [(1 - hA/2](1+Ah/2)^{-1} \tag{5.12}$$

Then:

$$B(h)[1 + hB(h)]^{-1} = B(h) = A[1 - Ah/2]^{-1}[(1 - hA/2](1+Ah/2)^{-1}$$

$$= A[1 - hA/2][1 - A^2 h^2/4]^{-1} \tag{5.13}$$

Using 4.9, 5.12 and 5.13:

$$[1+ hB(h)]^n = [1 + hA/2]^n(1-Ah/2)^{-n}$$

$$= \exp(t\int_{[0, 1]} A[1 - uhA/2][1- A^2u^2h^2/4]^{-1}du) \quad (5.14)$$

5.11 now follows from 5.14.

q.e.d.

Theorem 5.3. The centered-difference approximation to $\{t \to \exp(tA)\}$ leads to a power series represention in the step-size h which only contains even powers of h:

Proof. This is a consequence of a symmetry property in 'h' of the centered difference approximation, namely:

The matrix valued function $\{h \to [1 - Ah/2]^{-1/h}[1 + Ah/2]^{1/h}\}$ (5.11) is invariant under the transformation: $\{h \to -h\}$.

Since the function $\{h \to [1 - Ah/2]^{-1/h}[1 + Ah/2]^{1/h}\}$ is real-analytic in h, 5.11 implies that its power series expansion in 'h' contains only even powers.

q.e.d.

6. L. Richardson's 'deferred approach to the limit'.

Let us follow the beginning of Richardson's Great Paper [5]. Richardson's Method, in its classical form, can be exhibited in the form of the following result:

Theorem 6.1. Suppose given a real-valued function $\{h \rightarrow F(h)\}$ of a real variable h of the following form:

$$F(h) = a_0 + a_1 h^p + G(h) h^q, \qquad (6.1)$$

where $\{a_0, a_1\}$ are real numbers and $\{h \rightarrow G(h)\}$ is a real-valued function. Let λ be a real number such that: $0 < \lambda < 1$. Set:

$$F_1(h) = F(h) + \left[F(h) - F(\lambda h)\right]\left[\lambda^p - 1\right]^{-1} \qquad (6.2)$$

$$G_1(h) = G(h) - [1 - \lambda^p]^{-1}[G(h) - \lambda^q G(\lambda h)] \qquad (6.3)$$

Then:
$$F_1(h) = a_0 + G_1(h) h^q \qquad (6.4)$$

Proof. $F(h) - F(\lambda h) = a_0 + a_1 h^p + G(h) h^q - a_0 - a_1 \lambda^p h^p - \lambda^q G(\lambda h) h^q$

$$= a_1 h^p [1 - \lambda^p] + h^q [G(h) - \lambda^q G(\lambda h)] \qquad (6.5)$$

Hence:

$$F_1(h) = a_0 + a_1 h^p + G(h) h^q -$$

$$[1 - \lambda^p]^{-1}\left[a_1 h^p [1 - \lambda^p] + h^q [G(h) - \lambda^q G(\lambda h)]\right]$$

$$= a_0 + h^q \left[G(h) - [1 - \lambda^p]^{-1}[G(h) - \lambda^q G(\lambda h)]\right] \qquad (6.6)$$

6.4 follows from 6.6.

q.e.d.

Corollary to Theorem 6.1. If $0 < p < q$, then, as $h \to 0$, $F_1(h)$ is a better approximation to a_0 than $F(h)$. Iterating the construction, we obtain a sequence:

$$\{F(h), F_1(h), \ldots\} \tag{6.7}$$

of better and better approximations to a_0. This sequence of approximations is the essence of the Richardson "deferred approach to the limit" method [5].

This method for improving approximations has many variations. Its widest use seems to be in constructing algorithms for ODE numerical solutions. Suppose that:

$$x_{app}(t, h) = x(t) + x_1(t)h + x_2(t)h^2 + \ldots \tag{6.8}$$

is the result of applying one of the one-step numerical methods for approximating the solution: $\{t \to x(t)\}$ of an ODE of the form 1.1. Let:

$$h_1, h_2 \text{ be two arbitrary real numbers,} \tag{6.9}$$
$$\text{in the range of the expansion 6.8}$$

Then:

$$x_{app}(t, h_1) = x(t) + x_1(t)h_1 + x_2(t)h_1^2 + \ldots \tag{6.10}$$

$$x_{app}(t, h_2) = x(t) + x_1(t)h_2 + x_2(t)h_2^2 + \ldots \tag{6.11}$$

Multiply 6.10 by h_2 and 6.11 by h_1 and subtract:

$$h_2 x_{app}(t, h_1) - h_1 x_{app}(t, h_2) = x(t)(h_2 - h_1) + x_2(t)[h_2 h_1^2 - h_1 h_2^2] \ldots$$

hence:

$$[h_2 x(t, h_1) - h_1 x(t, h_2)]/(h_2 - h_1) = x(t) + x_2(t)[h_2 h_1^2 - h_1 h_2^2]/(h_2 - h_1) \ldots \tag{6.12}$$

This identity can be interpreted as follows:

As $h_1, h_2 \to 0$, for fixed t, the vector:

$$[h_2 \aleph_{app}(t, h_1) - h_1 \aleph_{app}(t, h_2)]/(h_2 - h_1) \qquad (6.13)$$

approximates $\aleph(t)$ to a higher order than did $\{\aleph_{app}(t, h)\}$

Here is another interpretation of 6.13. Set:

$$h_2 = h \text{ and } h_1 = \lambda h \qquad (6.14)$$

$$y(t, h) = [\aleph_{app}(t, \lambda h) - \lambda \aleph_{app}(t, h)]/(1 - \lambda) \qquad (6.15)$$

Then:

$\{y(t, h)\}$ is a 'better' approximation to $\{\aleph(t)\}$ than $\{\aleph_{app}(t, h)\}$ $\qquad (6.16)$

Notice that:

$\{y(t, h)\}$ is the result of transforming $\{\aleph_{app}(t, h)\}$ in an $\qquad (6.17)$
algebraic (and group-theoretic/vector bundle-theoretic) way.

6.18 again can then be iterated to get a sequence of better-and-better approximations to $\{t \to \aleph(t)\}$.

Remark. This iteration process has some of the flavor and essence of 'renormalization' and the 'renormalization group' that is so beloved nowadays by physicists and the pop-chaoticians!

7. Systems of difference equations which deform Frobenius Integrable Differential Systems.

Part of my Grand Design (begun in [3]) is to develop a Theory of Recursive Systems to parallel the Theory of Differential Systems. One reason that I concentrate here on the Numerical Analysis of ODE's is that this is the simplest playground for working out the ideas. From my point of view, Numerical Analysis of ODE's (or PDE's for that matter) involves:

Systems of Difference Equations
+
A deformation theory relating 'difference' to 'differential' equations.

In this Section, I will illustrate this Principle further by discussing the approximation by systems of difference equations of Frobenius Integrable PDE Systems of type 1.4-1.5.

Let us first discretize the system 1.4-1.5 in the following way: Introduce a System of Partial Difference Equations:

$$\aleph_{app}(s, t+h) = \aleph_{app}(s, t) + hF(\aleph_{app}(s, t), s, t, h, k) \qquad (7.1)$$

$$\aleph_{app}(s+k, t) = \aleph_{app}(s, t) + kG(\aleph_{app}(s, t), s, t, h, k). \qquad (7.2)$$

where:

$$h, k \; \varepsilon \; (-a, a), \text{ for some positive real number } a. \qquad (7.3)$$

F, G are smooth maps: $\qquad (7.4)$
$$R^n \times R^2 \times (-a, a) \times (-a, a) \longrightarrow R^n$$

Definition. The Recursive System 7.1-7.2 is said to be a **deformation** of the PDE System 1.4-1.5 iff:

$$F(x, s, t, 0, 0) = f(x, s, t); \; G(x, s, t, 0, 0) = g(x, s, t) \qquad (7.5)$$
for all $x \; \varepsilon \; R^n$.

Again, let us look to the linear case for further guidance.

8. Difference System approximations for linear Frobenius Integrable PDE systems in two independent variables.

Introduce the following Lie-group theoretic notation:

$$GL(n, R) = \text{the Lie group of nxn real matrices.} \qquad (8.1)$$

$$\mathbf{GL(n, R)} = \text{the Lie algebra of } GL(n, R) \qquad (8.2)$$

$$= \text{the Lie algebra of nxn real matrices, with the Lie bracket the commutator of two matrices.} \qquad (8.3)$$

Let us suppose given the following data:

$$\mathbf{x} \in \mathbf{GL(n, R)}; \ s, t \in \mathbf{R} \qquad (8.4)$$

Two smooth maps $\{(s, t) \rightarrow A(s, t), C(s, t)\}$ of $\mathbf{R}^2 \rightarrow \mathbf{GL(n, R)}$, $\qquad (8.5)$

Associate with the data 8.4-8.5 the following PDE system:

$$\mathbf{x}_t = A(s, t)\mathbf{x} \qquad (8.6)$$

$$\mathbf{x}_s = C(s, t)\mathbf{x}, \qquad (8.7)$$

to be solved for a map:

$$(s, t) \rightarrow \mathbf{x}(s, t) \qquad (8.8)$$
$$\mathbf{R}^2 \rightarrow \mathbf{GL(n, R)}$$

The Frobenius Integrability condition for this system is the following:

$$A_s - C_t = [A, C] \qquad (8.9)$$

For each matrix $\mathbf{x}_{(0, 0)}$ in $GL(n, R)$, the PDE system 8.3-8.4 then determines a smooth map:

$$\mathbb{R}^2 \longrightarrow GL(n\ R) \qquad (8.10)$$
$$(s, t) \longrightarrow \varkappa(s, t)$$

which is a solution of 8.3-8.4 and which satisfies the following initial condition:

$$\varkappa(0, 0) = \varkappa_{(0, 0)} \qquad (8.11)$$

Thus we face the Computer Science Problem of describing effectively and computationally this solution. This Problem is a simple prototype of many others encountered in Applications in Mechanics, Control, Elementary Particle Physics, etc.

The simplest case is that where the following conditions are satisfied:

$$A_s = A_t = C_t = C_s = [A, C] = 0 \qquad (8.12)$$

Theorem 8.1. Suppose that conditions 8.12 are satisfied. Given a solution $\{(s, t) \longrightarrow \varkappa(s, t)\}$ of the system 8.3-8.4, we can in this case write down an explicit formula for the solution:

$$\varkappa(s, t) = \exp(tA)(\varkappa(s, 0)) \qquad (8.13)$$

$$\varkappa(s, 0) = \exp(sC)(\varkappa(0, 0)) \qquad (8.14)$$

$$\varkappa(s, t) = \exp(tA)\exp(sC)(\varkappa(0, 0))) \qquad (8.15)$$

Proof. Obvious.

The form 8.15 of the solution enables us to write down approximate solutions which are solutions of difference equations, using the methods developed in Sections 2-6. I will first do this the traditional way, using finite-difference approximations for the partial derivatives. Then, I will describe more Lie-theoretic approaches.

9. Difference equation approximations to a linear Frobenius System.

As we have done for linear difference equations, we can construct discrete versions of the PDE's 8.3 and 8.4. We use either of the following two approximations:

$$\varkappa_t(s, t) \approx [\varkappa(s, t+h) - \varkappa(s, t)]/h \qquad (9.1)$$

$$\varkappa_s(s, t) \approx [\varkappa(s+k, t) - \varkappa(s, t)]/k \qquad (9.2)$$

or

$$\varkappa_t(s, t) \approx [\varkappa(s, t+h/2) - \varkappa(s, t-h/2)]/h \qquad (9.3)$$

$$\varkappa_s(s, t) \approx [\varkappa(s+k/2, t) - \varkappa(s-k/2, t)]/k \qquad (9.4)$$

9.1-9.2 is called the **forward difference approximation**, and 9.3-9.4 the **centered difference approximation**. The approximation 9.1-9.2 leads to the following system of difference equations:

$$[\varkappa_{app}(s, t+h) - \varkappa_{app}(s, t)]/h = A(s, t)\varkappa_{app}(s, t) \qquad (9.5)$$

$$[\varkappa_{app}(s+k, t) - \varkappa_{app}(s, t)]/k = C(s, t)\varkappa_{app}(s, t) \qquad (9.6)$$

The approximation 9.3-9.4 leads to the following system of difference equations:

$$[\varkappa_{app}'(s, t+h/2) - \varkappa_{app}'(s, t-h/2)]/h = \\ [A(s, t+h/2)\varkappa_{app}'(s, t+h/2) + A(s, t-h/2)\varkappa_{app}'(s, t-h/2)]/2 \qquad (9.7)$$

$$[\varkappa_{app}'(s+k/2, t) - \varkappa_{app}'(s-k/2, t)]/k = \\ [C(s+k/2, t)\varkappa_{app}'(s+k/2, t) + C(s-k/2, t)\varkappa_{app}'(s-k/2, t-h/2)]/2 \qquad (9.8)$$

Let us rewrite 9.5-9.6 as follows:

$$\aleph_{app}(s, t+h) = [hA(s, t) + 1]\aleph_{app}(s, t) \tag{9.9}$$

$$\aleph_{app}(s+k, t) = [kC(s, t) + 1]\aleph_{app}(s, t) \tag{9.10}$$

We can solve 9.9-9.10 as follows:

$$\aleph_{app}(s, h) = [hA(s, 0) + 1]\aleph_{app}(s, 0) \tag{9.11}$$

$$\aleph_{app}(s, 2h) = [hA(s, h) + 1][hA(s, 0) + 1]\aleph_{app}(s, 0) \tag{9.12}$$

$$\cdots$$

$$\aleph_{app}(p, 0) = [kC(0, 0) + 1]\aleph(0, 0) \tag{9.13}$$

$$\aleph_{app}(2p, 0) = [kC(p, 0) + 1][kC(0, 0) + 1]\aleph(0, 0) \tag{9.14}$$

$$\cdots$$

$$\aleph_{app}(pk, nh) = \{\,\}\,\aleph(0, 0) \tag{9.15}$$

where the terms in the brackets { } involve 'time-ordered' products of the matrices $[hA(s, t)+1]$ and $[kC(s, t)+1]$

Solutions analogous to 9.15 can be written down for the centered-difference approximate difference equations 9.7-9.8. In either case, these approximate solutions are combinatorially extremely complicated, particularly when one wants to exhibit the dependence on 'h' and 'k'. (In fact, the physicists have developed an elaborate combinatorial methodology - called 'Feynman Diagrams' - to keep track of such computations!) In order to find situations where they can be considered more explicitly (other than the situation $A_s = A_t = C_t = C_s = [A, C] = 0$ considered in Section 8) I will, in the following Sections, consider two situations where more explicit infomation can be obtained.

10. Separable Frobenius Integrable Linear PDE Systems

Consider again the PDE system:

$$X_t = A(s, t)X \tag{10.1}$$

$$X_s = C(s, t)X, \tag{10.2}$$

with the following Frobenius Integrability Condition:

$$A_s - C_t = [A, C] \tag{10.3}$$

Definition. The system 10.1-10.2 is said to be **separable** if the following conditions are satisfied:

The coefficient function $\{A(s, t)\}$ is a function $\{A(s)\}$ of 's' alone (10.4)

The coefficient function $\{C(s, t)\}$ is a function $\{C(t)\}$ of 't' alone. (10.5)

If 10.4-10.5 is satisfied, then, using 10.3, 'A' and 'C' satisfy the following differential equations:

$$dA/ds - dC/dt = -[C(t), A(s)] \tag{10.6}$$

Theorem 10.1. One solution of equations 10.3-10.6 is the following:

$$dA/ds = 0; \tag{10.7}$$

$$C(t) = \exp(Ad(-At))(C(0) = C(0) - t[A, C(0)] + t^2/2[A,[A, C(0)]] - \ldots \tag{10.8}$$

Proof. The hypothesis 10.7 leads - via 10.6 - to the differential equation:

$$dC/dt = [C(t), A(s)] \tag{10.9}$$

10.7 and 10.4 imply that 'A' is a constant matrix. 10.9 then becomes a differential equation for {t --> C(t)}, of which 10.8 is the general solution.

q.e.d.

Returning to the case where 10.8 is not necessarily satisfied, we have:

Theorem 10.2. The General Solution of the PDE system 10.1-10.5 can be written in the following form:

$$\aleph(s, t) = \exp(tA(s))\exp(sC(t))(\aleph(0, 0))) \tag{10.10}$$

Proof. Similiar to that of 8.15.

Theorem 10.3. If one-discrete parameter semi-group approximations of the form $\{m \to (1+hB(h, s))^m\}$ and $\{p \to (1+kD(k, t)^p\}$ are used for the one-parameter groups $\{t \to \exp(tA(s))\}$ and $\{s \to \exp(sC(t))\}$ (as in Section 4), then the following formula provides approximations to 10.10:

$$\aleph_{app}(s, t; h, k) = (1+hB(h, s))^{t/h}(1+kD(k, t))^{s/k}(\aleph(0, 0))) \tag{10.11}$$

Proof. Use 4.9 in 10.10.

Let us analyze further the conditions that equations 10.4-10.6 impose:

Theorem 10.4. Conditions 10.4-10.6 imply the following differential equations for $\{s \to A(s)\}$ and $\{t \to C(s)\}$:

$$d^2A/ds^2 = [C(t), dA/ds] \tag{10.12}$$

$$d^2C/dt^2 = [A(s), dC/dt] \tag{10.13}$$

$$[dC(t)/dt, dA/ds] = 0 \tag{10.14}$$

Proof. 10.12 and 10.13 result from differentiating 10.6 with respect to 's' and 't', respectively. 10.14 now follows on differentiating 10.12 with respect to 't'.

q.e.d.

Let us now consider further assumptions about the system 10.1-10.3 which will enable us to write down a General Solution which can be approximated in a form where the dependence on the step-size parameters can be made more explicit.

Theorem 10.4. Conditions 10.4-10.6 imply the following explicit formulas for $\{s \rightarrow A(s)\}$ and $\{t \rightarrow C(s)\}$:

$$dA/ds = Ad(\exp(C(t))[dA/ds](0)] \qquad (10.15)$$

$$dC/dt = -Ad(\exp(A(s))[dC/dt](0)] \qquad (10.16)$$

$$A(s) = sAd(\exp(C(t))[dA/ds](0)] + A(0) \qquad (10.17)$$

$$C(t) = -tAd(\exp(A(s))[dC/dt](0)] + C(0) \qquad (10.18)$$

Proof. 10.15 and 10.16 result from solving the linear ODE's 10.12 and 10.13 for $\{s \rightarrow dA/ds\}$ and $\{t \rightarrow dA/dt\}$, respectively. 10.17 and 10.18 result from solving 10.15 and 10.16, respectively.

q.e.d.

11. Frobenius Integrable Linear PDE Systems associated with a smooth family of 2-dimensional Lie subalgebra of GL(n, R).

Consider again the Frobenius Integrability equation 8.9:

$$A_s - C_t = [A, C] \qquad (11.1)$$

Suppose that the following additional condition is satisfied:

There are real numbers a and c such that:

$$[A(s, t), C(s, t)] = aA(s, t) + cC(s, t) \text{ for all } (s, t) \qquad (11.2)$$

Remark. Lie algebraically, 11.2 says that:

For each (s, t), A(s, t)} and C(s, t) generate a (11.3)
2-dimensional Lie subalgebra of the Lie algebra **GL(n, R)**.

This subalgebra may 'move' with (s, t), i.e. we are presented with a **deformation of Lie subalgebras of the Lie algebra GL(n, R) [4]**.

11.1 then takes the following form:

$$A_s - C_t = aA + cC \qquad (11.4)$$

Let us again suppose that:

The system 10.1-10.2 is separable in the sense (11.5)
that 10.4-10.5 is satisfied.

Then, we have:

The differential system 11.4 is equivalent to the following system of ODE's:

$$dA/ds - aA(s) = B = dC/dt + cC(t) \qquad (11.6)$$
where 'B' is a constant nxn matrix.

We can of course solve the ODE system 11.6 as follows:

$$A(s) = \exp(as)[A(0) + B/a] - B/a \qquad (11.7)$$

$$C(t) = \exp(-ct)[C(0) - B/c] + B/c \qquad (11.8)$$

We now have:

Theorem 11.1. The General Solution of the PDE system 10.1-10.5 can be written in the following form:

$$\aleph(s, t) = \exp(tA(s))\exp(sC(t))(\aleph(0, 0))) \tag{11.9}$$

with 11.7-11.8 satisfied.

12. Frobenius Integrable Linear Differential Equations and their discrete approximations associated with two dimensional Lie subalgebras of **GL(n, R)**.

Let us now specialize further the hypotheses of Sections 10 and 11. Suppose that A and C are two fixed, linearly independent elements of **GL(n, R)** satisfying the following commutation relation:

$$[A, C] = aA + cC, \tag{12.1}$$

where a and c are fixed real numbers. (Lie algebraically, 12.1 then means simply that {A, C} generate a 2-dimensional Lie subalgebra of **GL(n, R)**.

Given:
$$\aleph_{(0, 0)} \in \mathbf{GL(n, R)}, \tag{12.2}$$

set:
$$\aleph(s, t) = \exp(tA)\exp(sC)\aleph_{(0, 0)}. \tag{12.3}$$

Let us now ask the following question:

> Which linear PDE systems are satisfied by (12.4)
> the collection of maps $\{(s, t) \to \aleph(s, t)\}$,
> where '$\aleph_{(0, 0)}$' may be any element of **GL(n, R)**?

Let us tackle this by differentiating 12.3:

$$ℵ_t(s, t) = A\exp(tA)\exp(sC)ℵ_{(0, 0)} = Aℵ(s, t) \tag{12.5}$$

$$ℵ_s(s, t) = \exp(tA)C\exp(sC)ℵ_{(0, 0)} \tag{12.6}$$

Using the commutation relation 12.1, we have:

$$\exp(tA)C = [\exp(tA)C\exp(-tA)]\exp(tA) \tag{12.7}$$

$$\exp(tA)C\exp(-tA) = Ad(\exp(tA))(C)$$

$$= \exp(tAdA)(C)$$

$$= C + t[A, C] + t^2 1/2[A, [A, C]] + \ldots$$

$$= C + t(aA + cC) + t^2[1/2c(aA + cC)] + \ldots \tag{12.8}$$

Set:
$$C(t) = \text{right hand side of 11.8.} \tag{12.9}$$

Then, we have:

$$ℵ_s(s, t) = C(t)ℵ_s(s, t) \tag{12.10}$$

Let us sum up as follows:

Theorem 12.1. Let $\{(s, t) \to ℵ(s, t)\}$ be defined by 12.3. Then, for every set of initial conditions $\{ℵ_{(0, 0)}\}$, $\{(s, t) \to ℵ(s, t)\}$ satisfies the following PDE system:

$$ℵ_t(s, t) = Aℵ(s, t) \tag{12.11}$$

$$ℵ_s(s, t) = C(t)ℵ_s(s, t) \tag{12.12}$$

Proof. Combine 12.5 and 12.10.

Theorem 12.2. The General Solution of the linear PDE system 12.11-12.12, with given initial conditions, can be written in the following form:

$$\aleph(s, t) = \exp(tA)\exp(sC)\aleph(0, 0) \qquad (12.13)$$

Proof. Follows from Theorem 12.1 and 12.3.

Theorem 12.3. Here are approximations for the General Solution 12.13 of the PDE system 12.11-12.12:

$$\aleph_{app}(s, t; h, k) = (1+hB(h, s))^{t/h}(1+kD(k, t))^{s/k}(\aleph(0, 0)), \qquad (12.14)$$

where:

$$\exp(tA) \approx (1+hB(h, s))^{t/h} \qquad (12.15)$$

$$\exp(tC) \approx (1+hD(h, s))^{t/h} \qquad (12.16)$$

For example, the Euler Approximation is:

$$\aleph_{app}(s, t; h, k) = (1+htA)^{t/h}(1+ksC)^{s/k}(\aleph(0, 0)), \qquad (12.14)$$

Proof. 12.14 follows from 12.3 and 12.15-12.16.

I will now study further the method of using approximations of the General Solution in order to construct approximating systems of difference equations.

13. Difference equations obtained from approximations of the General Solution and commutation relations.

Let us again study PDE systems whose General Solution can be written in the following form:

$$\aleph(s, t) = \exp(tA)\exp(sC)(\aleph(0, 0))) \qquad (13.1)$$

Let us use the approximations given in Section 4 for the matrix-valued exponential functions on the right-hand side of 13.1:

$$\exp(tA) \approx (1 + hB(h))^{t/h} \qquad (13.2)$$

$$\exp(sC) \approx (1 + kD(k))^{s/k} \qquad (13.3)$$

where:

$$\{h \longrightarrow B(h)\} \text{ and } \{k \longrightarrow D(k)\} \text{ are nxn matrix-valued functions.} \qquad (13.4)$$

Insert these approximations into 13.1, obtaining:

$$\aleph_{app}(s, t; h, k) = (1 + hB(h))^{t/h}(1 + kD(k))^{s/k}(\aleph(0, 0))) \qquad (13.5)$$

Theorem 13.1 Suppose that, for fixed (h, k), the matrices B(h) and D(k) satisfy the following commutation relations:

$$hk[B(h), D(k)] = \alpha(h, k)(1 + hB(h)) \qquad (13.6)$$

where $\alpha(h, k)$ is a real number. Set:

$$m = t/h; \; p = s/k \qquad (13.7)$$

Then, for fixed (h, k), the map: $\{(m, p) \longrightarrow \aleph_{app}(pk, mh; h, k)\}$ (where $\{\aleph_{app}(s, t; h, k)\}$ is given by 13.5) is a solution of the following system of difference equations:

$$\aleph_{app}(m+1, p) = (1 + hB(h))\aleph_{app}(m, p) \qquad (13.8)$$

$$\aleph_{app}(m, p+1) = \big[m\alpha(h, k) + (1 + kD(k))\big]\aleph_{app}(m, p) \qquad (13.9)$$

Proof. Insert 13.7 into 13.5 and use relation 13.6.

Remark. Condition 13.6 is again an example of a **Deformation of Lie Algebra Structure [2, 4]**. I will now approach more systematically the topic of approximating an example of a PDE system from the point of view of Lie Algebra Deformation Theory [2, 5] and Lattice Gauge Theory [6].

14. Finite Difference Approximations to Frobenius Integrable Linear PDE's obtained by deforming the connection and curvature operator of a vector bundle.

Consider again a linear, time-dependent PDE system satisfying the Frobenius Integrability Condition:

$$ℵ_t(s, t) = A(s, t)ℵ(s, t) \tag{14.1}$$

$$ℵ_s(s, t) = C(s, t)ℵ(s, t) \tag{14.2}$$

$$A_s(s, t) - C(s, t)_s = [A(s, t), C(s, t)] \tag{14.3}$$

In order to apply more systematically the ideas of Deformation Theory, let us recast the PDE system 14.1-14.2 into the following framework:

Set:
Γ = real vector space of smooth mappings: $\quad(14.4)$
$\gamma : \{(s, t) \to ℵ(s, t)\}$ of $\mathbb{R}^2 \to \mathbb{R}^n$

$\partial_t, \partial_s: \Gamma \to \Gamma$ are the linear operators of $\quad(14.5)$
partial derivative.

$\mathbb{A}, \mathbb{C}: \Gamma \to \Gamma$ are the linear operators defined as follows:

$\mathbb{A}\gamma : \{(s, t) \to ℵ(s, t)\} \to \{(s, t) \to A(s, t)ℵ(s, t)\} \quad(14.6)$

$\mathbb{C}\gamma : \{(s, t) \to ℵ(s, t)\} \to \{(s, t) \to C(s, t)ℵ(s, t)\} \quad(14.7)$

We then have the following commutation relations among the linear operators on Γ defined above:

$$[\partial_t, \mathbf{A}]: \{(s, t) \to \mathbf{x}(s, t)\}$$

$$= \{(s, t) \to \partial_t(A(s, t)\mathbf{x}(s, t)) - A(s, t)\partial_t\mathbf{x}(s, t)\}$$

$$= \{(s, t) \to \partial_t(A(s, t))\mathbf{x}(s, t))\} \tag{14.8}$$

We can abbreviate the relation 14.8 as follows:

$$[\partial_t, \mathbf{A}] = \partial_t(\mathbf{A}) \tag{14.9}$$

Similarly, let us set:

$$[\mathbf{A}, \mathbf{C}] = \{(s, t) \to [A(s, t), C(s, t)]. \tag{14.10}$$

Theorem 14.1. With these notations, the PDE system 14.1-14.2 can be written in the following form:

$$\left[\partial_t - \mathbf{A}\right](\gamma) = 0 \tag{14.11}$$

$$\left[\partial_s - \mathbf{C}\right](\gamma) = 0 \tag{14.12}$$

$$[\partial_s, \mathbf{A}] - [\partial_t, \mathbf{C}] - [\mathbf{A}, \mathbf{C}] = 0. \tag{14.13}$$

The left hand side of 14.3 can be defined more algebraically as follows:

$$[\partial_t - \mathbf{A}, \partial_s - \mathbf{C}] = \partial_s(\mathbf{A}) - \partial_t(\mathbf{C}) + [\mathbf{A}, \mathbf{C}] \tag{14.14}$$

Proof. Follows from the formulas 14.3, 14.6, 14.7 and 14.10.

Let us now introduce the Forward Difference Operators as follows:

$$\Delta_h: \{(s, t) \to \aleph(s, t)\} \to \{(s, t) \to [\aleph(s, t+h) - \aleph(s, t)]/h\} \quad (14.15)$$

$$\Delta_k: \{(s, t) \to \aleph(s, t)\} \to \{(s, t) \to [\aleph(s+k, t) - \aleph(s, t)]/k\} \quad (14.16)$$

For fixed (h, k), let us replace the system with a system of the following form:

$$[\Delta_h - A(h, k)](\gamma) = 0 \quad (14.17)$$

$$[\Delta_k - C(h, k)](\gamma) = 0 \quad (14.18)$$

where:

$A(h, k)$ and $C(h, k)$ are linear maps: $\Gamma \to \Gamma$. $\quad (14.19)$

Definition. The system of difference equations 14.17-14.18 is said to satisfy the **Frobenius Integrability Condition** if the following condition is satisfied:

$$[\Delta_h - A(h, k), \Delta_k - C(h, k)] = 0. \quad (14.20)$$

Theorem 14.2. The Integrability Condition 14.20 is equivalent to the following condition:

$$-[\Delta_h, C(h, k)] + [\Delta_k, A(h, k)] + [A(h, k), C(h, k)] = 0. \quad (14.21)$$

Proof. 14.21 follows from 14.20 and the following relation:

$$[\Delta_h, \Delta_k] = 0. \quad (14.22)$$

Thus, we can summarize as follows:

Theorem 14.3. The construction of a linear system of difference equations approximating a linear Frobenius PDE system can be interpreted as finding families:

$$\{(h, k) \longrightarrow A(h, k), C(h, k)\} \tag{14.23}$$

satisfying the relations 14.21 and the following conditions:

$$A(0, 0) = A; \ C(0, 0) = C \tag{14.24}$$

Remark. What I have done in this Section can also be interpretd in language similiar to that used by physicists in Lattice Gauge Theory. Geometrically, this requires that the vector space Γ be identified with the space of smooth cross-sections of a vector bundle whose base is R^2. A and C define a linear connection in this bundle. $\partial_t - A$ and $\partial_s - C$ are operators of covariant differentiation associated with this connection. They satisfy the commutation relations 14.14, i.e.:

$$[\partial_t - A, \partial_s - C] = \partial_s(A) - \partial_t(C) + [A, C] \tag{14.25}$$

Geometrically, the right hand side of 14.25 defines the **curvature** of the connection. Thus, the Frobenuis Integrability Condition requires the vanishing of this curvature. (In Yang-Mills Theories one imposes more general conditions on this curvature.) Lattice Gauge Theory then proceeds to study difference equation approximation to the resulting PDE systems constructed by means analogous to those invoved in Theorem 14.2.

15. The Frobenius Integrability Condition for systems of difference equations.

Let us now briefly return to a study of (possibly) nonlinear systems of difference equations. For simplicity, I will only study a system consisting of two difference equations, of the following form:

$$x(p, m+1) = F(x(p, m), p, m) \tag{15.1}$$

$$x(p+1, m) = G(x(p, m), p, m) \tag{15.2}$$

$$x \in R^n; \ p, m \in I = \text{non-negative integers} \tag{15.3}$$

$$F \text{ and } G \text{ are maps: } R^n \times I \times I \longrightarrow R^n \tag{15.4}$$

Let us investigate the conditions of compatibility between 15.1 and 15.2:

$$x(p+1, m+1) = F(x(p+1, m), p+1, m) = F(G(x(p, m), p, m), p+1, m) \tag{15.5}$$

$$x(p+1, m+1) = G(x(p, m+1), p, m+1) = G(F(x(p, m), p, m), p, m+1) \tag{15.6}$$

Definition. The system 15.1-15.2 of difference equations is said to be **Frobenuis Integrable** if the following condition is satisfied:

$$F(G(x, p, m), p+1, m) = F(G(x, p, m), p+1, m) \tag{15.7}$$
$$\text{for all } x \in R^n; \ p, m \in I$$

Theorem 15.1. If 15.7 is satisfied, and $x_{(0, 0)}$ is a fixed element of R, then the system 15.1-15.2 has a unique solution $\{(p, m) \longrightarrow x(p, m)\}$ such that:

$$x(0, 0) = x_{(0, 0)} \tag{15.8}$$

Proof. We can find a map: $\{(0, m) \rightarrow \aleph(0, m)\}$ such that:

$$\aleph(0, m+1) = F(\aleph(0, m), 0, m) \tag{15.9}$$

and

$$\aleph(0, 0) = \aleph_{(0, 0)}. \tag{15.10}$$

We can then find a map: $\{(p, m) \rightarrow \aleph(p, m)\}$ such that:

$$\aleph(p+1, m) = G(\aleph(p, m), p, m) \tag{15.11}$$

$$\{(0, m) \rightarrow \aleph(0, m)\} \text{ is the solution of 15.9-15.10.} \tag{15.12}$$

It only remains to prove that 15.2 is satisfied: This follows from the Frobenius Integrability Condition 15.7 by reversing the argument given above in 15.5-15.6.

q.e.d.

Remark. No doubt these computations can be interpreted in terms of a 'non-linear covariant derivative' associated to a 'connection'. I plan to study this in a later work extending the ideas in this paper

16. The nonlinear ODE one-step convergence theorem from the point of view of the theory of transformation groups.

As we have seen, in the case of linear systems of ODE's or PDE's of a certain type the approximations involved in the standard numerical analysis discretization process can be understood from the point of view of the Lie group $GL(n, R)$ and its sub-semigroups. It is then mathematically natural to try to study the non-linear cases 1.1 or 1.4-1.5 by extending the ideas to the group of diffeomorphisms of R^n, and possibly more general spaces and mappings. I now begin this process.

Suppose given the following data:

A topological space $\mathcal{H}$ (16.1)

$M(\mathcal{H})$ = semigroup (under mapping composition) of continuous maps: $\mathcal{H} \to \mathcal{H}$ (16.2)

$(-a, a) = \{h \in \mathbf{R}: -a < h < a\}$ is an interval of real numbers containing the origin.   (16.3)

$\phi: \mathcal{H} \times (-a, a) \to \mathcal{H}$, $\{(x, h) \to \phi(x, h)\}$, is a continuous map.   (16.4)

For $h \in (-a, a)$, $\phi_h: \mathcal{H} \to \mathcal{H}$ is the map: $\{x \to \phi(x, h)\}$ (16.5)

For each $t \in \mathbf{R}$, $h \in (-a, a)$, $h = t, t/2, \ldots$ set:

$$\alpha(t, h) = [\phi_h]^{t/h} \qquad (16.6)$$

We ask:

Does there exist a map $\beta: \mathbf{R} \times \mathcal{H} \to \mathcal{H}$ such that:

$$\beta(t, x) = \lim_{h \to 0} \alpha(t, h)(x) \text{ for } (t, x) \in \mathbf{R} \times \mathcal{H}? \qquad (16.7)$$

Remark. This is just a general setting for the convergence-of-discrete approximations of ODE's proved in the linear case in earlier sections and for nonlinear situations on Euclidean spaces in the numerical analysis literature, e.g. Gear's book [1], Theorem 4.1.

Let us now examine under what conditions the limit map β defines a semigroup. Suppose the map β satisfying 16.7 exists. Apply the 'currying' process to it; for each $t \in \mathbf{R}$, let:

$$\beta_t: \mathcal{H} \to \mathcal{H} \qquad (16.8)$$

be the map defined by:

Lie-Theoretic Numerical

$$\beta_t(x) = \beta(t, x) \text{ for } x \in \mathbf{H}. \tag{16.9}$$

Use 16.7:

$$\beta(t+t', x) = \lim_{h \to 0} \alpha(t+t', h)(x) =, \text{ using 16.6,}$$

$$\lim_{h \to 0} [\phi_h]^{(t+t')/h}(x) =$$

$$\lim_{h \to 0} [\phi_h]^{t/h}\left([\phi_h]^{t'/h}(x)\right) \tag{16.13}$$

$$\beta(t, \beta(t', x)) = \lim_{h \to 0} [\phi_h]^{t/h}(\beta(t', x))$$

$$= \lim_{h \to 0} [\phi_h]^{t/h}\left(\lim_{h' \to 0} [\phi_{h'}]^{t'/h'}(x)\right) \tag{16.14}$$

We have then proved:

Theorem 16.1. For $t, t' \in \mathbf{R}$, we have:

$$\beta_{t+t'} = \beta_t \beta_{t'}, \tag{16.10}$$

if and only if the following condition is satisfied:

$$\lim_{h \to 0} [\phi_h]^{t/h}\left([\phi_h]^{t'/h}(x)\right) = \lim_{h \to 0} [\phi_h]^{t/h}\left(\lim_{h' \to 0} [\phi_{h'}]^{t'/h'}(x)\right) \tag{16.13}$$

Proof. Use 16.7:

$$\beta(t+t', x) = \lim_{h \to 0} \alpha(t+t', h)(x) =, \text{ using 16.6,}$$

$$\lim_{h \to 0} [\phi_h]^{(t+t')/h}(x) =$$

$$\lim_{h \to 0} [\phi_h]^{t/h}\left([\phi_h]^{t'/h}(x)\right) \tag{16.13}$$

$$\beta(t, \beta(t', x)) = \lim_{h \to 0} [\phi_h]^{t/h}(\beta(t', x))$$

$$= \lim_{h \to 0} [\phi_h]^{t/h}\left(\lim_{h' \to 0} [\phi_{h'}]^{t/h'}(x)\right) \tag{16.14}$$

Remark. This work poses the research problem of studying sufficient conditions (at the level of generality 16.1-16.2) for 16.14 to be satisfied. Further, one might ask for a 'Lie Theory' of such relations, extending the Lie Theory which is traditional for continuous-time dynamical systems on manifolds. I leave such a study for a later work

17. Conditions for the existence of a limiting smooth dynamical system for deformations of discrete semigroups on manifolds.

Let us now add to the assumptions of Section 16 the following hypotheses:

$\mathbf{H}$ is a smooth (i.e. infinitely differentiable) manifold.

$$\phi: \mathbf{H} \times (-a, a) \to \mathbf{H}, \{(x, h) \to \phi(x, h)\}, \text{ is a smooth map.} \tag{17.1}$$

$$\phi(x, 0) = x \text{ for all } x \in \mathbf{H} \tag{17.2}$$

For each $x \in \mathbf{H}$, set:

$$V(x) = \text{tangent vector to the curve } \{h \to \phi(x, h)\} \text{ at } h=0. \tag{17.3}$$

Theorem 17.1. The map V: {x --> V(x)} defines a smooth vector field on K.

Proof. Follows immediately from 17.1-17.3.

Theorem 17.2. Suppose that K is the manifold R^n. Identify a tangent vector at each point of K with K itself. Then, there is a smooth map

$$f: K \times (-a, a) \rightarrow K \qquad (17.4)$$

such that:

$$\phi(x, h) = x + hf(x, h) \qquad (17.5)$$

$$f(x, 0) = V(x) \text{ for all } x \in K. \qquad (17.6)$$

Proof. Since ϕ is, by hypothesis, infinitely differentiable as a function of both 'x' and 'h', a Taylor expansion the map {K --> ϕ(K, h)} about the point (x, o) gives a representation of the form 17.5. (We use condition 17.2 to guarantee that the first term on the right hand side of 17.6 is as stated.) 17.6 now follows from 17.3.

q.e.d.

Theorem 17.3. Suppose that:

$$x_0 \text{ is a point of } \mathbf{X} \qquad (17.7)$$

$$\mathbf{U} \text{ is a coordinate neighborhood of } \mathbf{X} \qquad (17.8)$$

such that:

$$x_0 \in \mathbf{U} \qquad (17.9)$$

For all non-negative integers $\{n\}$, all h in (-a, a),
$$\phi(x, h)^n \in \mathbf{U} \qquad (17.10)$$

Suppose that:

$\{t \longrightarrow x(t)\}$ is the orbit curve of the vector field V beginning at x for 't = 0', i.e. the solution of the ODE:

$$dx/dt = V(x) \qquad (17.11)$$
such that:
$$x(0) = x_0. \qquad (17.12)$$

Then, for all positive real numbers 't' such the solution of 17.11-17.12 exists and is in **U** for the interval [0, t], we have:

$$x(t) = \lim_{h \to 0} [\phi_h]^{t/h}(x_0) \qquad (17.13)$$

Proof. Our hypotheses guarantee that Theorem 4.1 of Gear's book [1] applies.

18. Final remarks about future research directions.

I believe that the ideas outlined in this paper suggest much further development. Here are four research directions which I expect to be fruitful:

a). Develop a Theory of Discrete Appoximations for more of the PDE systems which, in classical terms, may be solved using ODE algorithms.

b). Develop the Deformation Theory (and associated cohomology) which is appropriate to the numerical solution of differential systems.

c). Develop Algorithms for the solution of the approximations of PDE systems which are adapted to Parallelism

d). Develop the relations to the Theory of Connections and to Lattice Gauge Theory.

Bibliography

1. W. Gear, **Numerical Initial Value Problems in Ordinary Differential Equations**, Prentice-Hall, 1971

2. M. Hazewinkel and M. Gerstenhaber, **Deformation Theory of Algebras and Structures and Applications**, Kluwer, 1988

3. R. Hermann, **Geometric Computing Science: First Steps**, Math Sci Press, Brookline, MA 1991

4. A. Nijenhuis and R. Richardson, Deformation of homomorphisms of Lie groups and Lie algebras, **Bull. Am. Math. Soc.**, 73, 175-179, 1967

5. L. F. Richardson, The deferred approach to the limit, **Trans. Roy. Soc. Lon.**, 226, 1927, 299-349

6. K. Wilson, The confinement of quarks, Phys. Rev. D, v. 10, 1974, 2445.

CHAPTER 2 - COMPUTING THE ORBITS OF LIE GROUP ACTIONS

1. Introduction.

As part of my interest in the interaction between 'pure' and 'applied' mathematics (which I call 'Interdisciplinary Mathematics'), I have long cultivated a dilletante's interest in the Numerical Analysis of Differential Equations. (For example, I contributed some ideas, explained in "Interdisciplinary Mathematics, vol. 21, about a Lie Group interpretation of some of the algorithms of matrix numerical analysis, which were carried to completion by G. Ammar and C. Martin) However, this interest became more focussed about five years ago, when I was writing Volume 25, titled "Geometric Computing Science". Motivated by Control Theory, I had the idea that there was a mathematical structure to the process of Modelling the 'real, continuous' world in terms of the 'discrete' world of the computer, and that this structure involved what geometers call **Deformation Theory.** (I had encountered Deformation Theory first-hand through my interaction with my thesis advisor, Don Spencer - who was one of the originators of the mathematics of Deformation Theory - and then through understanding how it might be 'applied' to Physics and Control Theory. I have also been continually inspired by the pioneering and seminal work in the 1960's by Nijenhuis and R. Richardson on Deformation Theory in a Lie-theoretic context.)

I then turned to Differential Equation Numerical Analysis as one of the areas where my general ideas could be put to the test. Gerry Sussman of MIT is the Computer Scientist with whom I have most extensively talked over these ideas. Reciprocally, I have learned much about Computer Science - and sharpened the insights I had already - from talking with him, his students and collaborators. For the past year, I have been a Research Affilliate in his Lab at MIT (supported by a 1-year Exploratory Research Grant from the Computational Mathematics Program of NSF), and have become interested in the research program on the Numerical Analysis of the long-time behavior of the ODE'S of Celestial Mechanics which he and his astrophysicist colleague Jack Wisdom have been pursuing. Gerry asked me a question about the geometric setting for the Richardson Extapolation Method of numerical solution of ODE's. Considering this question in the coordinate-free, manifold framework in which we differential geometers like to think led me to broaden the outlook to more general differential systems. Rather than pursue the path that a Computer Scientist or Applied Mathematician would find most natural, I plan to broaden the scope and study certain general geometric topics which I find most interesting. Based on my experiences in Mechanics, Control and Elementary Particle Physics, I claim that this intellectual approach of a Differential Geometer will find its uses in future Applications!

My Research involves examining the existing ODE Numerical Algorithms from a Lie-theoretic point of view, and then extending them to more general Lie groups, partially using 'deformation' ideas due to Kodaira and Spencer, Nijenhuis and R. Richardson. As Application, I will study certain situations in Celestial Mechanics Numerical Analysis (with particular attention to the 2- and restricted 3-Body Problem and associated Regularizations) which I believe can benefit from this Lie point of view. I will be particularly concerned with what the physicists call the **'Runge-Lenz Vector'** in Newtonian 2-particle dynamics (explained from the Lie-theoretic point of view in my books "Lie Groups for Physicists" and Volume 27, "Constrained Mechanics and Lie Theory"). I believe that the Lie Theory of this situation clarifies considerably the classical 2- and 3-Body Regularization Theory (presented in the book by Stiefel and Scheifele), and leads to many insights into the natural discretizations of the dynamic equations which may be appropriate for numerical purposes.

2. The geometric and deformation-theoretic setting for some ideas in ODE numerical analysis.

In the standard treatises, ODE Numerical Analysis is concerned with smooth systems of the following form:

$$dx/dt = f(x); x \; \varepsilon \; R^n. \tag{2.1}$$

Computing Orbits

The aim is to approximate at fixed time 't' the known (via the Picard ODE Existence Theorem) solution $\{t \longrightarrow x(t)\}$ of system 2.1 by solutions of a difference equation of the form:

$$x(n+1) = x(n) + hg(x(n), h), \qquad (2.2)$$

as the 'step-size' parameter 'h' goes to zero and n goes to infinty, in such a way that 'hn' remains constantly equal to 't'. A basic Numerical Analysis Existence Theorem (see Theorem 4.3 of Gear's treatise) states that a sufficient condition (in addition to 'smoothness') for such convergence is that:

$$g(x, 0) = f(x) \text{ for all } x \in R^n. \qquad (2.3)$$

In Volume 25, I have remarked that this condition has a 'deformation-theoretic' meaning, analogous to the geometric conditions found by Kodaira and Spencer in their famous work on Deformations of Complex Structures and Nijenhuis and Richardson in their work on Deformations of Lie Group Homorphisms.

I will now turn to the manifold setting. Let:

X be a smooth manifold

and let:

$$\{R \times X \longrightarrow X\}$$

be a smooth action of the Lie group of the additive real numbers R on X. Analytically, this can be represented by the following formulas:

$$\{(t, x) \longrightarrow \phi(t)(x)\},$$

where:

$\{t \longrightarrow \phi(t)\}$ is a homorphism of the Lie group R to the group **DIFF**(X) of diffeomorphisms of X.

$\{t \to \phi(t)\}$ has an infinitesimal generator, which is a smooth vector field V on X. We can dEscribe this relation in terms of the following formula:

$$\{\phi(t) = \exp(tV)\}.$$

Let **MAP**(X) be the semigroup of smooth maps of X to itself. ODE Numerical Analysis then involves a mapping:

$$\{R \to \mathbf{MAP}(X)\}, \{h \to \alpha(h)\}$$

such that:
$$\exp(tV) = \lim_{h \to 0} [\alpha(h)]^{t/h} \tag{2.4}$$

Relation 2.4 can be interpreted in deformation-of-group-homomorphism form, as developed in the work of Nijenhuis and Richardson. Namely:

$\{h \to \{t \to \alpha(h)]^{t/h}\}\}$ is a family (parameterized by 'h') (2.5)
of groups (or, more generally perhaps, semigroups) which
goes over, as $\{h \to 0\}$, to the group $\{t \to \exp(tV)\}$.

I will now generalize, replacing the abelian, 1-dimensional Lie Group R by more general objects.

3. Towards a Numerical Analysis of Lie group actions on manifolds.

Suppose given the following data:

$$G \text{ is a connected, finite dimensional Lie group.} \tag{3.1}$$

$$X \text{ is a smooth manifold.} \tag{3.2}$$

$$\mathbf{DIFF}(X) = \text{group of diffeomorphisms on X.} \tag{3.3}$$

A smooth transformation group action of G on X:

$$G \times X \longrightarrow X; \tag{3.4}$$
$$(g, x) \longrightarrow gx$$

Let:
$$\phi: G \longrightarrow \mathbf{DIFF}(X) \tag{3.5}$$

be the group-homomorphism defined as follows:

$$\phi(g)(x) = gx. \tag{3.6}$$

Now, the case 'G=R' corresponds to the classical ODE situation considered in Section 2.

Research Problem: Formulate a reasonable meaning to the concept of 'Numerical Analysis' of the homomorphism 3.5, generalizing the case 'G= R'. Provide, if possible, a connection with the Nijenhuis-R. Richardson Theory, generalizing the case 'G=R'. In particular, develop methods for computing the orbits of G acting on X, i.e. given a point $x_0 \in X$, 'compute' the subset:

$$\{\phi(g)(x_0) | \ g \ \varepsilon \ G\} \tag{3.7}$$

I will now suggest some methods for tackling this 'computing the G-orbit' problem. The first step is to define coordinate systems for the manifold structure on G which are naturally associated with the group structure. To this end, let:

$$\{t \longrightarrow g_1(t), ..., g_m(t)\} \text{ be a collection of} \tag{3.8}$$
$$\text{one-parameter subgroups of G.}$$

Assign to the data 3.8 a map:

$$\beta: R^m \longrightarrow G \tag{3.9}$$

constructed as follows:

$$\beta(t_1, ..., t_m) = g_1(t_1)\cdots g_m(t_m) \quad \text{for } (t_1, ..., t_m) \; \varepsilon \; R^m. \tag{3.10}$$

Definition. If the following condition is satisfied:

The map β is a local diffeomorphism in the neighborhood U (3.11)
of $(0, ..., 0)$ of R^m,

then the Cartesian coordinates defined on the open subset $\alpha^{-1}(U)$ of G are called the **canonical coordinates** for G associated with the data 3.8-3.10.

Remark. The classical **Euler Angle** parameterization of the rotation group $G = SO(3, R)$ is constructed in this way.

Consider the collection of one-parameter groups of diffeomorphisms of X:

$$\{t \dashrightarrow \phi(g_1(t)), ..., t \dashrightarrow \phi(g_m(t))\} \tag{3.12}$$

They have infinitesimal generators, which are a collection:

$$\{V_1, ..., V_m\} \tag{3.13}$$

of smooth vector fields on X. We have, for $x \; \varepsilon \; X$:

$$\phi(g_1(t))(x) = \exp(tV_1)(x), \;...,\; \phi(g_m(t))(x) = \exp(tV_m)(x) \tag{3.14}$$

Following the ODE Numerical Analysis ideas sketched in Section 2, we can 'aPproximate' the G-orbit as follows:

Choose a collection:

$$\{h \dashrightarrow \alpha_1(h), ..., \alpha_m(h)\} \tag{3.15}$$

of step-size dependent, smooth maps: $X \dashrightarrow X$, and approximate the one-parameter diffeomorphisms 3.12 as follows:

$$\phi(g_1(t))(x) \approx [\alpha_1(h)]^{t/h}(x)$$
$$\cdots \qquad (3.16)$$
$$\phi(g_m(t))(x) \approx [\alpha_m(h)]^{t/h}(x)$$

Later on in this book, I will turn to the Kepler Problem of Celestial Mechanics for another Example which motivates the search for a General Theory.

CHAPTER 3 - SOME GEOMETRIC PDE THEORY FOR SMOOTH MAPS

1. Introduction.

What is most exciting to me about the Colombeau-Oberguggenberger-Rosinger theory is that it suggests a re-thinking of the General Theory of Differential Systems from the 'infinitely differentiable' and 'real-analytic' categories where it originated to the 'generalized' situations which extend the Category of Smooth (or Real-Analytic) maps between smooth manifolds. In this Chapter, I will first review some of the traditional ideas (based on a paper and book of mine [6a, 12]), as preparation for extensions to Generalized Maps. and as a Geometric Setting for the Mechanics material in this Volume. I will also describe in this Chapter the Differential Calculus on Manifolds and Jet Bundle Calculus Notations used in this Volume.

When I was under Ehresmann's tutelage from 1953-55, he emphasized to me that he saw his Jet Calculus as inspired by the 'classical' theory of systems of partial differential equations. (I was very intrigued by the fact - which I learned much later - that Ehresmann was set on this course by contact with Vessiot at the Ecole Normale in the early 30's.) Unfortunately, this 'classical theory' is very difficult for a modern reader to understand, and no one in the 1950's put the effort into preparing an adequate expository account for the modern reader.

I started to write such a treatment in 1959 while I was an instructor at Harvard. Unfortunately, I could not get the support I needed to carry it to completion at that time. What I **did** do had a certain resonance in today's scientific world: I took a job at Lincoln Laboratory and utilized the work that I had already done on interpreting Lie, Vessiot, etc. to prepare a modern exposition of Mechanics, which I elaborated on in Lectures and Lecture Notes at Berkeley in1961-62, and which in turn was used by Ralph Abraham in preparing his "Foundations of Mechanics". I also used this material in work on nonlinear control systems and the Lagrange Problem in the Calculus of Variations, which ten years later was picked up by others (some of whom were not even aware of the extent of my work!) and led to what we call today "Geometric Control Theory".

In the past five years, my thoughts have returned to this material for various reasons. First, I believe that fundamental progress in what physicists ambitiously call Nonlinear Physics depends on their utilizing the geometric insights of the sort of PDE theory pioneered by Lie and Cartan, as they absorbed some other work of these Masters for their triumphs in Particle Physics. As I mentioned already in my Polemics in Vol. 21, I wonder at the solidity of the currently accepted foundations of Particle Physics, particularly without a more profound understanding of the Nonlinear PDE foundations (and not to speak of such subjects as Turbulence!). I remain a Skeptic that we **really** know what the hell an Interacting, Nonlinear Quantum Field Theory is! As I have already propagandized in Vol. 26, the Generalized Function Algebra tools of Colombeau, Oberguggenberger and Rosinger give us marvelous new mathematical insights, which have yet to be aborbed into the Nonlinear Physics and Quantum Field Theoretic (or the Nonlinear PDE) cultures.

Remark. By the way, I don't think that this C-O-R Mathematics supports 'Einstein's Dream'of a Magic Set of Equations Which Will Explain All! It might be more like Logic: A Set of Axioms, with many Models, which might change dramatically as shorter distances are probed. Of course, even such an outcome would be a triumph of the human intellect on the scale of what Uncle Albert tried to do!

Second, the subject of Generalized Function Algebras and its Interaction with nonlinear PDE theory is, in my opinion, sufficiently fascinating to grab the attention of the world's researchers in nonlinear PDE for the next twenty years or so, just as the Schwartz distribution theory dominated the thoughts of researchers in linear PDE theory in the period 1960-1980. (I am amazed that those I have spoken to who consider themselves to be the 'leaders' do not see to be aware of this!) I also see the Colombeau-Oberguggenberger-Rosinger theory fitting in very well with insights of Geometric PDE Theory which we have inherited from Lie and Cartan.

Geometric PDE Theory

The material in this Chapter is based partially on a paper [6a] I wrote about ten years ago, as well as my book [9], and does not directly involve the C-O-R Generalized Functions. However it deals with classical material which will ultimately extend in some form to the C-O-R context, and will also have important Applied conext. See also the material in "Geometry, Physics and Systems" [9] for other material of this flavor

The theory of Differential Systems depends on a confusing mixture of 'classical' material (some of which is listed in the Bibliography of this Chapter) and 'modern' concepts. I have been especially influenced by those 'modern' concepts developed by my Teachers, Charles Ehresmann and Don Spencer, and those 'classical' concepts learned from my extensive study of Sophus Lie, Elie Cartan, Goursat, Vessiot, ... In the limited space of this Chapter, I of course cannot go into full detail; I just cover enough of the Foundational material to describe some Classical topics which have interested me since my graduate student and post-doc studies of the 1950's.

2. Some concepts from jet space theory.

I will now review the Ehresmann jet space ideas, in order to set the stage.

All manifolds will be C^∞, finite-dimensional, and paracompact, unless mentioned otherwise. All maps between manifolds will also be "smooth," i.e., C^∞. Let X and Y be such manifolds. A **local map from X to Y** is a pair:

$$(U, \phi) \qquad (2.1)$$

consisting of:

$$\text{an open subset U of X} \qquad (2.2)$$

and:

$$\text{a } C^\infty \text{ map } \phi: U \to Y. \qquad (2.3)$$

Geometric PDE Theory

Let $\mathbf{M}(X,Y)$ be the space of all triples of the form:

$$(x, U, \phi) \tag{2.4}$$

where:

$$U \text{ is an open subset of } X, \tag{2.5}$$

$$\phi: U \to Y \text{ is a smooth map}, \tag{2.6}$$

$$x \in U. \tag{2.7}$$

Construct maps:

$$\mathbf{s}: \mathbf{M}(X, Y) \to X \tag{2.8}$$

$$\mathbf{t}: \mathbf{M}(X, Y) \to Y \tag{2.9}$$

as follows:

$$\mathbf{s}(x, \phi, U) = x, \tag{2.10}$$

$$\mathbf{t}(x, \phi, U) = \phi(x). \tag{2.11}$$

s is called the **source map**, and **t** the **target map**.

We will construct the spaces that are basic to the Ehresmann theory as quotients of $\mathbf{M}(X, Y)$ under equivalence relations. The source and target maps **s** and **t** will be constant on these equivalence classes, and hence will pass to the quotient to define maps from the space of equivalence classes to X and Y, respectively. For the sake of notational simplicity, we will use the same letter to denote these maps on the equivalence classes.

Remark. The Colombeau-Oberguggenberger-Rosinger Theory also involves equivalence relations on functions spaces. I foresee a fruitful field of Future Research to work out the general settings for such equivalence gadgets!

Geometric PDE Theory

First, consider the following equivalence relation:

$$(x, \phi, U) \sim (x', \phi', U') \tag{2.12}$$

iff:

$$x = x' \tag{2.13}$$

and

$\phi = \phi'$ on an open subset of $U \cap U'$
containing x. (2.14)

The quotient of **M**(X, Y) under the equivalence relation (2.12-.14) (i.e., the space of all equivalence classes) is called the **sheaf** of all maps: $X \to Y$, denoted as:

SM (X, Y).

For each integer r, we define another equivalence relation as follows:

$(x, \phi, U) \sim (x', \phi', U')$
iff $x = x'$, ϕ and ϕ' **meet to order r at x**, in the sense that, in a local coordinate system for X about x, the partial derivatives at x of ϕ and ϕ' agree to order r. (2.15)

The quotient of M(X, Y) by this equivalence relation is called the space of **rth-order jets** of maps $X \to Y$ denoted as:

$$J^r(X, Y). \tag{2.16}$$

Geometric PDE Theory

Theorem 2.1. The $J^r(X, Y)$,

$$r = 0, 1, 2, \ldots, \tag{2.17}$$

are C^∞, finite-dimensional, paracompact manifolds. The source and target maps:

$$\mathbf{s}: J^r(X, Y) \to X, \tag{2.18}$$

$$\mathbf{t}: J^r(X, Y) \to Y, \tag{2.19}$$

are submersions (i.e., the induced linear mapping on tangent bundles is onto), and hence define the Ehresmann jet spaces as fiber spaces.

Proof. Left to the reader.

Consider the following equivalence relation:

$$(x, \phi, U) \sim (x', \phi', U') \tag{2.20}$$

iff $x = x'$, ϕ and ϕ' meet to **infinite order** at x, in the sense that the Taylor expansions about x of ϕ and ϕ' in a local coordinate system agree.

Definition. The quotient of $\mathbf{M}(X, Y)$ by this equivalence relation is denoted by $J^\infty(X, Y)$, and called the space of **infinite-order jets**.

There are inclusions among the equivalence classes of these equivalence relations. For example:

Two maps ϕ and ϕ' which agree to rth order at x, agree to order s, where $s < r$.

This leads to natural maps between the various quotient spaces:

Geometric PDE Theory

$$\mathbf{SM}(X, Y) \to J^\infty(X, Y)$$
$$\ldots \to J^r(X, Y) \to J^{r-1}(X, Y) \to \ldots \to J^0(X, Y) = X \times Y. \tag{2.21}$$

Using the source and target maps, respectively, we obtain fiber space maps:

$$J^r(X, Y) \to X \tag{2.22}$$

$$J^r(X, Y) \to Y. \tag{2.23}$$

Differential forms on X and Y can be lifted up to $J^r(X, Y)$ via the pullback of the maps (2.22-2.23). For simplicity:

> **I often make no notational distinction between differential forms on X and Y and their pullback to $J^r(X,Y)$.**

If $\phi: U \to Y$ is a local map from X to Y, its **r-jet** or **prolongation** is the map:

$$j^r(\phi): U \to J^r(X, Y) \tag{2.24}$$

obtained as follows:

> For $x \in U$, construct the element (x, ϕ) of $\mathbf{M}(X, Y)$ and then assign it the equivalence class on $J^r(X, Y)$ to which it belongs. This is defined as $j^r(\phi)(x)$. As x varies over U, the map (2.7) is obtained.

We can now state what has been done from the point of view of Category Theory.

Geometric PDE Theory

Definitions. Let X and Y be finite dimensional, infinitely differentiable, paracompact manifolds. Let $\Gamma(X, Y)$ be the set of smooth (i.e. infinitely-differentiable maps) from X to Y. Let **N** be the following category:

$\qquad$ The objects of **N** are the non-negative integers N $\hfill$ (2.25)

$\qquad$ The morphisms of **N** are the pairs $(j, k) \in N \times N$, $\hfill$ (2.26)
$\qquad$ such that: $j \leq k$.

Let

$\qquad$ **DF** (which stands for 'differentiable fiber') $\hfill$ (2.27)
$\qquad$ be the category whose objects are the
$\qquad$ finite dimensional, infinitely-differentiable,
$\qquad$ paracompact manifolds and whose morphisms
$\qquad$ are the infinitely differentiable, local-product fiber maps
$\qquad$ between such manifolds.

Theorem 2.2. (Ehresmann) Let X and Y be finite dimensional, infinitely-differentiable, paracompact manifolds. There is a contravariant functor:

$$J(X, Y): \mathbf{N} \longrightarrow \mathbf{DF} \qquad (2.28)$$

$J(X, Y)$ assigns to each object j of the category **N** a manifold $J^j(X, Y)$, called the **manifold of j-th order jets of mappings** of X to Y, and it assigns to each morphism (j, k) of the category N, a local-product, smooth, fiber map:

$$\pi(k, j): J^k(X, Y) \longrightarrow J^j(X, Y), \qquad (2.29)$$

such that, whenever $j \leq k$ and $k \leq l$,

$$\pi(l, j) = \pi(k, j)\pi(l, k) \qquad (2.30)$$

(Theorem 2.2 continued on next page.)

Geometric PDE Theory

Further, there is a space $J^\infty(X, Y)$, called the space of **infinite order jets**, and a family of maps:

$$\pi(j): J^\infty(X, Y) \longrightarrow J^j(X, Y) \tag{2.31}$$

such that, whenever $j \leq k$,

$$\pi(j) = \pi(k, j)\pi(k) \tag{2.32}$$

For each $\gamma \in \Gamma(X, Y)$, each $k \in N$, there is a map:

$$j^k(\gamma): X \longrightarrow J^k(X, Y) \tag{2.33}$$

called the **prolongation** or **k-jet** of γ. The following condition is satisfied:

$$\pi(k, j)j^k(\gamma) = j^j(\gamma) \quad \text{for } j \leq k \tag{2.34}$$

In the language of Category Theory, $J(X, Y)$ is a contravariant functor from the category N associated with the ordered set of non-negative integers to the category **SET** of sets and maps. The collection of maps $\{\pi(j)\}$ defines $J^\infty(X, Y)$ as a 'colimit' of this functor.

Proof. Theorem 2.2 is a restatement of known (or obvious) properties of the Ehresmann jet space construction.

The bundle $J^\infty(X, Y)$ is algebraically constructed from the family:

$$\{J^k(X, Y): k \in N\}$$

of bundles. $J^\infty(X, Y)$ is 'infinite dimensional', while each $J^k(X, Y)$ is a finite dimensional, smooth manifold. The 'analysis' on $J^\infty(X, Y)$ is essentially equivalent to the theory of 'formal power series', and is rarely treated in the contemporary literature of 'geometric analysis'. I have a soft spot in my heart for this topic, since it was the subject of my first paper, called "Sur les jets d'ordre infinis", which I wrote at Ehresmann's suggestion and which he published in his 'Colloque de Topologie de Strasbourg' in 1954.

(What I did there turned out to be equivalent to a classical theorem of Emile Borel.)

3. Differential Systems and Integrability in the sense of Spencer and Goldschmidt.

The most important feature of Ehresmann's Jet Bundle Formalism is that it is the appropriate 'language' in which to express, in coordinate-free way, the Geometric Theory of Partial Differential Equations, which is of course so fundamental for the physical, biological and computational sciences.

We can now present the definition of a 'differential system':

Definitions. A subset **DS** $\subset J^k(X, Y)$ is said to define a **differential system of order k of maps from X to Y**. A $\gamma \in \Gamma(X, Y)$ is said to be a **solution** of **DS** if the following condition is satisfied:

$$\text{For each } x \in X, \; j^k(\gamma)(x) \in \mathbf{DS} \tag{3.1}$$

The collection of such solutions is denoted as:

$$\Gamma(\mathbf{DS}) \tag{3.2}$$

An element

$$\gamma \in \mathbf{DS} \cap (\text{fiber of } J^k(X, Y) \text{ above the point } x \in X) \tag{3.3}$$

is said to be **integrable** if there is a $\gamma \in \Gamma(\mathbf{DS})$ such that:

$$j^k(\gamma)(x) = \gamma \tag{3.4}$$

Remark. A main purpose of the work of Spencer [16a, 23 and Goldschmidt [5a] is to systematically study such Integrability properties. Spencer's main concern was the problem of studying 'integrability conditions' for systems of partial differential equations associated with the Lie-Cartan theory of pseudogroups, and certain of its subsystems. Cartan's

Geometric PDE Theory

theory of Exterior Differential Systems [1, 2, 14, 30] offers an alternate geometric formalism for many of these situations.

However, I will not attempt to expound and develop Spencer's extensive work at this point. It remains a Challenge to the mathematical world to Pick Up The Torch Passed on to Us by Don and carry forward the study of the Differential Systems and their Deformations that he pioneered. I believe that the Generalized Function tools developed by Colombeau, Oberguggenberger and Rosinger provide us with powerful new technique to attack the complex of geometric and analytic problems pioneered by Don.

4. DIFFERENTIAL GRADED ALGEBRAS AND THEIR DERIVATIONS.

One of the most remarkable features of Cartan's work on Differential Systems is how well he anticipated contemporary algebraic formalisms! Since he used coordinate-independent differential forms as far as possible, his work can often be readily algebracized. I tried my own hand at this, in Vol. 16, "Quantum and Fermion Geometry", which, written in the middle seventies, anticipated what happened (without any references to my work!) in the 1980's under the slogan "Non-Commutative Differential Geometry." There remains much to be done along these lines, particularly in 'Generalized Function Algebra' and 'Numerical Analysis' directions. In this Section, I will present a few concepts which will be useful in algebracizing various geometric ideas in Differential System Theory.

Definition. A set **D** is a a **differential graded algebra** (over the real numbers) if it has the following algebraic structures:

D is a vector space over the real numbers. (4.1)

D has a graded vector space stucture
$\{\mathbf{D}^n: n = 0, 1, \ldots\}$ (4.2)

D has an associative-algebra structure:

$$(\theta, \theta') \longrightarrow \theta \wedge \theta', \tag{4.3}$$

satisfying:

$$\mathbf{D}^n \wedge \mathbf{D}^m \subset \mathbf{D}^{n+m} \tag{4.4}$$

For $n = m = 0$, this product defines a ring structure on $\mathbf{D}^0$. Following custom, this product is defined as:

$$(f, f') \longrightarrow ff' \quad \text{for } f, f' \; \varepsilon \; \mathbf{D}^0. \tag{4.5}$$

For $n = 0$, $m \; \varepsilon \; N$, this product defines $\mathbf{D}^m$ as a module over the ring $\mathbf{D}^0$, with the product defined as:

$$(f, \theta) = f\theta \quad \text{for } f \; \varepsilon \; \mathbf{D}^0, \; \theta \; \varepsilon \; \mathbf{D}^m \tag{4.6}$$

c). **D** is a **differential algebra**, i.e. has an operation:

$$d: \mathbf{D} \longrightarrow \mathbf{D}, \tag{4.7}$$

called **exterior derivative**, satisfying the following conditions:

$$d^2 = 0 \tag{4.8}$$

d is R- linear $\tag{4.9}$

d is a **graded-derivation** of $\{\mathbf{D}^n: n = 0, 1, \ldots\}$, i.e.

$$d(\theta \wedge \theta') = d\theta \wedge \theta' + (-1)^m \theta \wedge d\theta' \tag{4.10}$$

for $\theta \; \varepsilon \; \mathbf{D}^m$, $\theta' \; \varepsilon \; \mathbf{D}$

$$d(\mathbf{D}^m) \subset \mathbf{D}^{m+1} \tag{4.10a}$$

Geometric PDE Theory

Definition. Let $\mathbf{D} = \{\mathbf{D}^n : n \in N, d, \wedge\}$ be a graded differential algebra, as indicated above. A **vector field** (relative to $\mathbf{D}$) is a map:

$$V: \mathbf{D} \longrightarrow \mathbf{D} \tag{4.11}$$

satisfying the following conditions:

$$V \text{ is R-linear.} \tag{4.12}$$

$$V(\mathbf{D}^n) \subset \mathbf{D}^n \quad \text{for } n \in N \tag{4.13}$$

$$V(\theta \wedge \theta') = V\theta \wedge \theta' + \theta \wedge V(\theta') \text{ for } \theta, \theta' \in \mathbf{D}. \tag{4.14}$$

$$V(d\theta) = dV(\theta) \text{ for } \theta \in \mathbf{D} \tag{4.15}$$

There is an R - linear map $\qquad$ (4.16)

$$V \rfloor : \mathbf{D} \longrightarrow \mathbf{D} \tag{4.17}$$
$$\theta \longrightarrow V \rfloor \theta$$

such that:

$$V(\theta) = d(V \rfloor \theta) + V \rfloor d\theta \text{ for all } \theta \in \mathbf{D} \tag{4.18}$$

$$V \rfloor (\theta \wedge \theta') = (V \rfloor \theta) \wedge \theta' + (-1)^n \theta \wedge (V \rfloor \theta') \tag{4.19}$$
$$\text{for } \theta \in \mathbf{D}^n, \theta \in \mathbf{D}$$

$$V \rfloor \mathbf{D}^n \subset \mathbf{D}^{n-1} \tag{4.19a}$$

The operation $\theta \longrightarrow V \rfloor \theta$ is called the **inner product** or **contraction** operation.

Geometric PDE Theory

Remark. Note that I have not assumed that the operation '∧ ' is skew-commutative, as it is for the case of exterior multiplication. For example, this 'non-commutative' framework would apply to 'quantum' situations, as suggested in Volume 16.

Theorem 4.1. Let **D** be a differential graded algebra. Let V(**D**) be the collection of vector fields. Then:

$$V(\mathbf{D}) \text{ is a Lie algebra over the real numbers,} \quad (4.20)$$
with the Lie algebra operation [,] given by operator-commutator:

$$[V, V'](\theta) = VV'(\theta) - V'V(\theta) \quad (4.21)$$
$$\text{for } \theta \, \varepsilon \, \mathbf{D}, \, V, V' \, \varepsilon \, V(\mathbf{D})$$

$$V(\mathbf{D}) \text{ is a module over the ring } \mathbf{D}^0: \quad (4.22)$$

$$(f, V) \longrightarrow fV$$
$$(fV)(\theta) = f(V(\theta)). \text{ for } f \, \varepsilon \, \mathbf{D}^0, \, \theta \, \varepsilon \, \mathbf{D}$$

Proof. Direct verification, left to the reader.

Canonical example: The smooth differential forms on a finite dimensional, infinitely differentiable, paracompact manifold. Vector fields and Lie derivative.

Let X be a finite dimensional, infinitely differentiable, paracompact manifold. Then, it is a standard exercise in 'calculus on manifolds' to define a differential graded algebra:

$$\mathbf{D}(X) = \{\mathbf{D}^m(X): m = 0, 1, \ldots \}, \quad (4.23)$$

with the elements of $\mathbf{D}^m(X)$ identified with the infinitely differentiable differential forms of degree m. In differential geometric terms, $\mathbf{D}^m(X)$ can identified with the smooth cross-sections of the vector bundle on X whose fibers are the exterior product of m copies of the cotangent vector spaces to the manifold X. The operations 'd' and '∧' required to provide D(X) with a differential graded algebra structure - as defined above - are given by:

$$d = \text{exterior derivative operation} \qquad (4.24)$$

$$\wedge = \text{exterior multiplication operation} \qquad (4.25)$$

These operations were first codified in the work of Elie Cartan.

This 'geometric' realization of a differential graded algebra also has a Lie algebra attached. Set:

$$V(X) = \text{set of smooth vector fields on X.} \qquad (4.26)$$

Elements of V(X) are defined, geometrically, as smooth cross-sections of the tangent vector bundle T(X) to X. Each $V \in V(X)$ defines a map:

$$\begin{aligned} D(X) &\longrightarrow D(X), \\ \theta &\longrightarrow V(\theta) \end{aligned} \qquad (4.27)$$

called **Lie derivative**. V(X) has a real Lie algebra structure:

$$(V, V') \longrightarrow [V, V'], \qquad (4.28)$$

called the **Jacobi-Lie bracket**. These operations have the algebriac properties described above in 4.11-4.22.

5. Moving Coframes of Differential Forms on the Jet Bundles.

There is a natural construction which leads from a pair of bases of differential forms on X and Y to a basis of differential forms on $J^r(X,Y)$. I will now describe it after a brief recapitulation of some algebraic properties of vector fields and differential forms on jet spaces.

Geometric PDE Theory

Definition. A one-differential form θ on $J^r(X, Y)$ is a **contact form** if

$$(j^r\phi)d(\theta) = 0 \qquad (5.1)$$
$$\text{for each local map } \phi: U \to Y.$$

Let $\mathbf{C}^r(X, Y)$ be the $\mathbf{F}(J^r(X, Y))$-module of contact forms. Suppose

$$\dim X = n, \quad \dim Y = m. \qquad (5.2)$$

Choose the following ranges of indices:

$$1 \leq i, j \leq n, \qquad (5.3)$$

$$1 \leq a, b \leq m. \qquad (5.4)$$

Suppose that:

(θ^a) is a basis for the one-forms $\mathbf{D}^1(Y)$ $\qquad (5.5)$

("basis" means "as an $\mathbf{F}(Y)$-module")
and that:

(ω^i) is a basis for $\mathbf{D}^1(X)$. $\qquad (5.6)$

Remark. Following Cartan's terminology, we might call (θ^a) and (ω^i) **moving coframes** for X and Y, respectively. See the material in Volume 28 for a more extensive treatment of this point of view.

Consider these as 1-forms on $J^r(X, Y)$ pulled back via the projection map. Then, there are functions

$$(y_i^a, y_{i_1 i_2}^a, \ldots, y_{i_1 \ldots i_r}^a) \qquad (5.7)$$

on $J^r(X, Y)$ such that the forms

Geometric PDE Theory

$$\eta^a = \theta^a - \Sigma_i y_i^a \omega^i, \tag{5.8}$$

$$\eta_{i_1}^a = dy_{i_1}^a - \Sigma_{i_2} y_{i_1 i_2}^a \omega^{i_2}, \tag{5.9}$$

$$\vdots$$

$$\eta_{i_1 \ldots i_{r-1}}^a = dy_{i_1 \ldots i_{r-1}}^a - \Sigma_{i_r} y_{i_1 \ldots i_r}^a \omega^{i_r}$$

are a basis for $\mathbf{C}^r(X, Y)$. Also, the forms

$$\omega^i, \theta^a, dy_i^a, \ldots, dy_{i_1 \ldots i_r}^a \tag{5.10}$$

form a basis for 1-forms on $J^r(X, Y)$.

For simplicity, let assume that the moving frames:

$$(\omega^i), (\theta^a) \tag{5.11}$$

are differentials of coordinate systems. We suppose that X and Y have a coordinate system, labeled as follows:

$$x^i, y^a, \tag{5.12}$$

with

$$\omega^i = dx^i, \quad \theta^a = dy^a. \tag{5.13}$$

One then proves readily that the functions

$$\left(x^i, y^a, y_i^a, \ldots, y_{i_1 \ldots i_r}^a\right) \tag{5.14}$$

associated with these bases form a coordinate system for $J^r(X, Y)$.

Remark. A symbolic notation such as:

$$(x, y, \partial_x y), (x, y, \partial_{xx} y, \ldots) \tag{5.15}$$

is often useful.

6. DIFFERENTIAL EQUATIONS, OPERATORS, AND SYMBOLS

Let $X, Y, J^r(X, Y), r = 0, 1, 2, \ldots$, be as before. Work with a fixed coordinate system:

$$(x^i) \text{ and } (y^a) \text{ for } X \text{ and } Y, \tag{6.1}$$

and the associated coordinates:

$$(x^i, y^a, y^a_i, \ldots) \text{ for } J^r(X, Y). \tag{6.2}$$

Definition: Let Z be a set.. An **rth-order differential operator symbol** with values in Z is a map:

$$\sigma: J^r(X, Y) \to Z. \tag{6.3}$$

Given such a symbol, we will define a map

$$D_\sigma: M(X, Y) \to M(X, V), \tag{6.4}$$

which is the associated **differential operator**.

For the following data:

$$\phi: U \to Y, \tag{6.5}$$

Geometric PDE Theory

$$U \text{ an open subset of } X, \qquad (6.6)$$

$$x \in U, \qquad (6.7)$$

D_σ is defined by:

$$D_\sigma(\phi)(x) = \sigma(j^r(\phi)(x)). \qquad (6.8)$$

One characteristic feature of this geometric approach to the theory of partial differential equations is the distinction between the "symbol," as a **geometric object**, i.e., as a mapping on a finite-dimensional manifold sitting above the domain and range manifolds X and Y ("independent" and "dependent" variables), and as a **mapping** between the infinite-dimensional space of local mappings. This correspondence is also implicit (and, in Lie's work, often quite explicit) in the 19th century literature on partial differential equations.

Definition. Let '0' denote a distinguished element of Z. The **differential equation** associated with such a differential operator is the subset

$$DE \equiv \sigma^{-1}(0) \qquad (6.9)$$

of $J^r(X, Y)$. The **solutions** of the differential equations are the $\phi \in M(X, Y)$ such that

$$D_\sigma(\phi) = 0, \qquad (6.10)$$

or, equivalently,

$$j^r(\phi)(U) \subset DE(\sigma). \qquad (6.11)$$

Remark: This geometric way of defining the geometric concept of "differential equation" can be defined in many different categories, e.g., the category of C^∞ maps, or algebraic maps, or, even with a little ingenuity, over "characteristic p." For the purpose of differential geometry and physics, the most important categories seem, at the present time at least, to be the C^∞ and the real or complex analytic maps.

7. PROLONGATIONS OF DIFFERENTIAL OPERATORS AND DIFFERENTIAL EQUATIONS

In the classical literature on nonlinear partial differential equations the process of **prolongation** is central, although not precisely defined. In Goldschmidt's and Spencer's work [6, 19, 26] one finds that it is an operator taking objects that live on a jet space of given order to one of next higher order. In this section I will describe a variant of their ideas.

With the notation explained in Section , let:

$$\sigma: J^r(X, Y) \to Z \tag{7.1}$$

be a symbol map. Assume that:

$$Z \text{ is a finite dimensional real vector space.} \tag{7.1a}$$

A local mapping $\phi: X \to Y$, expressed in local coordinates:

$$(x, y(x)) \text{ for } X, Y, \tag{7.2}$$

is then a **solution** of the differential equation associated with the symbol if:

$$\sigma(x, y(x), \partial_x y(x)) = 0 \quad \text{for all } x \in U. \tag{7.3}$$

Now, for fixed i, $1 \le i \le n$, differentiate (7.3) with respect to x^i. An $(r + 1)$th-order differential equation is obtained for which ϕ is a solution.

Geometric PDE Theory

The symbol of these prolonged differential equations, which will be a map denoted as

$$\delta_i \sigma: J^{r+1}(X, Y) \to Z, \tag{7.4}$$

will be a map:

$$J^{r+1}(X, Y) \dashrightarrow Z. \tag{7.5}$$

For example, if $r = 1$, and if σ is a function of the coordinates

$$(x^i, y^a, y_i^a) \tag{7.6}$$

on $J^1(X, Y)$, then:

$$\delta_i \sigma (x^i, y^a, y_i^a, y_{ij}^a) = \frac{\partial \sigma}{\partial x^i} + \Sigma_a \frac{\partial \sigma}{\partial y^a} y_i^a + \Sigma_{aj} \frac{\partial \sigma}{\partial y_j^a} y_{ij}^a. \tag{7.7}$$

Remark: Notice in formula (7.7) that the "affine" structure of the jet spaces described and utilized by Goldschmidt [5a] makes its appearance.

8. Prolongation of vector fields.

Keep the notations of Section 7. We can now describe the underlying geometric ideas in a more algebraic way by introducing the space of smooth vector fields $\mathbf{U}(X)$ on X. Suppose:

$$V = \Sigma_i A^i(x) \frac{\partial}{\partial x^i} \tag{8.1}$$

is such a vector field. Then, set

$$L_V(\sigma) = \Sigma_i A^i \delta_i(\sigma). \tag{8.2}$$

Geometric PDE Theory

Let us sum up as follows:

Theorem 8.1: For each integer $r \geq 0$, let $\mathbf{M}(J^r(X, Y), Z)$ be the space of C^∞ maps: $J^r(X, Y) \to Z$. [Algebraically, $\mathbf{M}(J^r(X, Y), Z)$ is $\mathbf{F}(J^r(X, Y)) \otimes Z$.] Thus, the prolongation described above is an **R**-linear map

$$L_V: \mathbf{M}(J^r(X, Y), Z) \to \mathbf{M}(J^{r+1}(X, Y), Z). \tag{8.3}$$

L_V, as the notation indicates, is a generalization of the usual Lie derivative.

The following formula holds:

For each map $\gamma: X \to Y$
$$L_V(j^r(\gamma)^*(\sigma)) = j^{r+1}(\gamma)^*(L_V(\sigma)). \tag{8.4}$$

Let:

$$t \to \exp(tV)$$

be the one-parameter pseudogroup on X generated by the vector field V. Then, for $x \in X$,

$$L_V(j^r(\gamma)^*(\sigma))(x) = \frac{\partial}{\partial t}(j^r(\gamma)^*(\sigma))[\exp(-tV)(x)]\Big|_{t=0}. \tag{8.5}$$

9. PROLONGATION OF DIFFERENTIAL EQUATIONS

Continue with the notation of Section 8. Let Z be a real vector space, and

$$\sigma: J^r(X, Y) \to Z$$

be an rth-order symbol map. The subspace

$$\sigma^{-1}(0) \equiv DE(\sigma) \tag{9.1}$$

of $J^r(X, Y)$ defines a **differential equation**. A map $\phi: X \to Y$ is a **solution** if the following condition is satisfied.

$$j^r(\phi)(X) \subset DE(\sigma). \tag{9.2}$$

An alternate way of putting this is that

$$j^r(\phi)^*(\sigma) = 0. \tag{9.3}$$

For each integer $s > 0$, and $V_1, \ldots, V_s \in V(X)$, consider the generalized Lie derivative

$$L_{V_1} \cdots L_{V_s}(\sigma). \tag{9.4}$$

It is a map $J^{r+s}(X, Y) \to W$. For each integer m, $m \geq s$, we can pull back the functions (9.4) to live on $J^{r+m}(X, Y)$, via the "forgetting" map: $J^{r+m} \to J^{r+s}$. In this way, we get a collection

$$\{\sigma, L_{V_1}(\sigma), \ldots, L_{V_1} \cdots L_{V_m}(\sigma): V_1, \ldots, V_m \in V(X)\} \tag{9.5}$$

of maps:

$$J^{r+m}(X, Y) \to W.$$

Let $DE^m(\sigma)$ be the subset of points of $J^{r+m}(X, Y)$ on which **all** the maps (9.5) **vanish**. Of course, as a differential equation, $DE^m(\sigma)$ has precisely the same set of solutions as $DE(\sigma)$. However, it is the geometric properties of each $DE^m(\sigma)$ that are of greatest interest.

In particular, one is interested in knowing something of the algebro-geometric-topological properties of $DE^m(\sigma)$ as subset of manifolds. Similar questions about subsets of jet spaces defined by equations occur in the theory of singularity of mappings. In this discipline, work can be done effectively in the category of C^∞ mapping; however, at the partial differential equation level, in the current state of mathematical development, serious work will require the assumption that the manifolds X, Y and the symbol map σ be **real analytic**.

Remark. One of the great Possibilities for the Theory of Differential Systems of the Colombeau-Oberguggenberger-Rosinger Theory is that it will enable one to transcend this 'smoothness' limitation!

When discussing the notion of **general** and **singular** solutions, later on in this paper, we shall make this assumption, since it will make available the powerful mathematical theorems to study these subsets that are not (yet) available in the C^∞ case. In fact, it is well known in the theory of partial differential equations that there are major qualitative differences between the C^∞ and real analytic category. If the coefficients of these equations were assumed analytic, the classic Cauchy-Kowalewsky theorem would guarantee the existence at least of local solutions.

Geometric PDE Theory

10. GENERAL SOLUTIONS OF SYSTEMS OF REAL-ANALYTIC PARTIAL DIFFERENTIAL EQUATIONS

The classical theory of 'General Solutions' of systems of differential equations is a good illustration of the capability of the Ehresmann Jet Bundle formalism for making precise what is very loosely stated in the classical literature.

Continue with the set up described in previous sections. Suppose given the following data:

$$X, Y, J^r(X, Y), \tag{10.1}$$

$$\sigma: J^r(X, Y) \to Z, \tag{10.2}$$

$$DE(\sigma) \subset J^r(X, Y), \tag{10.3}$$

the mth prolongation $DE^m(\sigma) \subset J^{r+m}(X, Y)$, for $m \geq 1$. $\tag{10.4}$

Now, assume that the data is **real analytic**.

$DE^m(\sigma)$ is a real analytic subvariety of the real analytic manifold $J^{r+m}(X, Y)$. $\tag{10.5}$

$DE^m(\sigma)$ is a union of a finite number of irreducible subvarieties. $\tag{10.6}$

Geometric PDE Theory

We can then consider the analytic real-valued functions:

$$f: U \to R \tag{10.7}$$

defined on open subsets U of $J^{r+m}(X, Y)$,

which satisfy the following condition:

f is identically zero on at least one irreducible branch of $DE^m(\sigma)$. (10.8)

Denote these functions by:

$$DE^m(\sigma, \mathbf{F}). \tag{10.9}$$

Remark. The form a sheaf. In general, one could make all of this much more precise by using sheaf-theoretic notations, ideas and terminolgy, but it would unduly complicate the exposition, particaluarly from the Applied point of view.

Let:

$\mathbf{GS}$ be a family of maps: $X \to Y$, (10.8)

such that:

each $\phi \in \mathbf{GS}$ is a solution of $DE(\sigma)$, i.e.

$$j^r(\phi)^*(\sigma) = 0. \tag{10.9}$$

Geometric PDE Theory

Then, for each integer:

$$m \geq 0, \tag{10.10}$$

and each:

$$\text{open subset } U \subset J^{r+m}(X, Y), \tag{10.11}$$

we can consider the set of maps:

$$f: U \to R \tag{10.12}$$

such that

$$J^r(\phi)^*(f) = 0 \quad \text{for all } \phi \in \mathbf{GS}. \tag{10.13}$$

Definition: **GS** is a **general solution** of the differential equations associated with the symbol map $\sigma: J^r(X, Y) \to Z$ if:

each function f satisfying (10.13), for an integer $m \geq 0$, belongs to $DE^m(\sigma, F)$. (10.14)

Remark. In words, condition 10.14 means that each differential equation satisfied by every map ϕ in the family GS is contained in the differential equation obtained by prolonging σ by differentiation.

This definition of "general solution" is one of two given and used in the 19th century literature [5, 6]; it is due to Ampère. The other is given Darboux' name, but we will not go into it here.

These generalities are best comprehended by thinking about certain simple examples.

11. THE GENERAL SOLUTION OF THE ONE-DIMENSIONAL WAVE EQUATION

Suppose given the following data:

$$X = R^2, \quad (11.1)$$
$$\text{with coordinates } (x^1, x^2),$$

$$Y = R, \quad (11.2)$$
$$\text{with coordinates } y.$$

$J^2(X, Y)$ has the following coordinates:

$$z = (x, y, t_1, y_2, y_{11}, y_{12}, y_{22}). \quad (11.3)$$

Define:

$$\sigma: J^2(X, Y) \to R \quad (11.4)$$

as follows:

$$\sigma(z) = y_{12}. \quad (11.5)$$

The solutions of the differential equations are the maps

$$x \to y(z) \quad (11.6)$$

such that

$$\frac{\partial y}{\partial x_1 \, \partial x_2} = 0. \quad (11.7)$$

This is (in "light cone" or "characteristic" coordinates) just the **Wave Equation**, in one space, one time variable.

To define the prolongations of (11.8), let:

Geometric PDE Theory

$$1 \le i, j, i_1, \ldots, \le 2 \tag{11.8}$$

be indices.

$$(x, y, y_i, y_{ij}, \ldots) \tag{11.9}$$

are coordinates of $J^\infty(X, Y)$.

The prolonged varieties are the subsets of $J^{r+m}(X, Y)$ obtained by setting equal to zero the following functions:

$$y_{i_1 \ldots i_s 12}, \tag{11.10}$$

where:

$$1 \le i_1, \ldots, i_s \le 2, \tag{11.11}$$

s is an arbitrary nonnegative integer.

Definition. The classical General Solution of the PDE (11.7) is:

The family of maps $R^2 \to R$, of the following form:

$$\phi(x_1, x_2) = g_1(x_1) + g_2(x_2), \tag{11.12}$$

where:

g_1, g_2 are arbitrary maps: $R \to R$. \hfill (11.13)

It is now obvious that any map

$$f: J^{r+m} \to R \tag{11.14}$$

which satisfies (11.2) is a function **only** of the variables (11.3). This is precisely the condition required to verify that the family of maps (11.4) is a general solution of the DE (11.2).

12. THE CLAIRAUT EQUATION: ITS "GENERAL" AND "SINGULAR" SOLUTION

The classical Clairaut equation is a first-order ordinary differential equation of the following form

$$y(x) - x\frac{dy}{dx} - f\left(\frac{dy}{dx}\right) = 0 \tag{12.1}$$

to be solved for a map:

$$x \to y(x) \tag{12.2}$$
$$\text{of } R \to R.$$

(f is a given C^∞ map: $R \to R$.)

To put this into the standard framework described above, let

$$X = Y = R. \tag{12.3}$$

Denoting:

$$\text{the coordinates of } J^1(X, Y) \text{ by } (x, y, y'), \tag{12.4}$$

the symbol σ is the map:

$$J^1(X, Y) \to R \tag{12.5}$$

defined as follows:

$$\sigma(x, y, y') = y - xy' - f(y'). \tag{12.6}$$

In order to define the prolonged systems, let us differentiate (12.1):

$$0 = \frac{dy}{dx} - \frac{dy}{dx} - x\frac{d^2y}{dx} - f'\left(\frac{dy}{dx}\right)\frac{d^2y}{dx^2}$$
$$= -\frac{d^2y}{dx^2}\left(x + f'\left(\frac{dy}{dx}\right)\right). \tag{12.7}$$

Thus,

$$\delta\sigma\left((x, y, y', y'')\right) = y''\left(x + f'(y')\right). \tag{12.8}$$

The variety of $J^2(X, Y)$ obtained by setting

$$0 = \sigma = \delta \tag{12.9}$$

is **reducible**.

At the differential equation level, C^∞ solutions of (12.1) are of two sorts:

Solutions of
$$\frac{d^2y}{dx^2} = 0 \tag{12.10}$$

or

$$f\left(\frac{dy}{dx}\right) = -x. \tag{12.11}$$

The General Solution of (12.10) is:

$$y = ax + b, \tag{12.12}$$

where a and b are real constants.

Geometric PDE Theory

The condition that 12.12 define a solution of (12.1) introduces the following relation between a and b:

$$ax + b - xa - f(a) = 0,$$

or

$$f(a) = b. \tag{12.13}$$

Theorem 12.1: The family

$$\phi_a(x) = ax + f(a), \tag{12.14}$$

parametrized by:

$$a \in R, \tag{12.15}$$

is a general solution, in the Ampère sense, of the differential equation (12.1).

Proof: Suppose:

$$\alpha(x, y, y', \ldots, y^{(r)}) \tag{12.16}$$

in an rth-order differential operator symbol, such that, for all a ε R:

$$y(x) = \phi_a(x) \tag{12.17}$$

, where '$\phi_a(x)$' is given by 12.14, is a solution of 12.1.

Geometric PDE Theory

Now,
$$\frac{d\phi_a}{dx} = a,$$

$$\frac{d^2\phi_a}{dx^2} = 0, \tag{12.18}$$

$$\vdots$$

Hence,
$$\alpha(x, ax + f(a), a, \ldots, 0) = 0, \tag{12.19}$$

and then
$$\alpha(x, y, z, \ldots, 0) = 0 \quad \text{for all } x, y. \tag{12.20}$$

We conclude that α must be of the following form:

$$\alpha = \beta_1 y'' + \ldots \beta_{r-2} y^{(r)}, \tag{12.21}$$

where:

$$\beta_1, \ldots, \beta_{r-2} \text{ are arbitrary functions on } J^r(X, Y). \tag{12.22}$$

This is what is required to prove the $\{\phi_a\}$ are "general solutions."

q.e.d.

The Clairaut equation is the example traditionally used in the classical treatises to introduce the notion of "singular solution." They are the solutions of the original differential equation (12.1), which are not part of the General Solution (12.2). They must then be solutions of (12.11). The general solution of (12.11) depends on one real parameter, say:

Geometric PDE Theory

$$x \to y^S(x; c) \tag{12.23}$$

with:

$$f'(y_x^S(x; c)) = -x. \tag{12.24}$$

Now, the condition that the family of functions of x, indicated in (12.23), **also** satisfy the original differential equation 12.1, is:

$$y^S(x, c) - xy_x^S - f(y_x^S) = 0. \tag{12.25}$$

In general, (12.24) and (12.25) will have solutions for only a discrete set of c's. These are the **singular solutions** of the Clairaut differential equation (12.1).

Now, the traditional geometric interpretation of these singular solutions of the Clairaut equation is that they are **the envelopes** of the family of curves in R^2 given by the General Solution:

$$x \to (x, y(x; a)) = (x, ax + f(a)). \tag{12.26}$$

For each a, (12.26) is a straight line. They are the tangent lines to the curves $x \to (x, y^S(x))$, where y^S are the solutions of (12.24) and (12.25).

13. COMPLETE SOLUTIONS OF PARTIAL DIFFERENTIAL EQUATIONS IN THE SENSE OF LAGRANGE AND VESSIOT. LAGRANGE-VESSIOT FOLIATIONS AND FIBRATIONS

In the classical literature, the concept of a "complete solution" of a partial differential equation is traditionally defined in the context of nonlinear first-order partial differential equations for one unknown function. The most important example for physical purposes (and the typical one from the mathematical point of view) is the Hamilton-Jacobi equation:

$$H\left(q, \frac{\partial S}{\partial q}\right) = 0, \qquad (13.1)$$

where H is a function of n "position" variables:

$$(q^1, \ldots, q^n), \qquad (13.2)$$

and "momentum" variables:

$$(p_1, \ldots, p_n). \qquad (13.3)$$

Remark. In more modern language, H is a real-valued function on the cotangent bundle $T^d(Q)$ to a manifold Q.

Geometric PDE Theory

Definition. A **complete solution** of (13.1) is a set of functions

$$q \to S(q; a), \quad (13.4)$$

parameterized by additional parameters:

$$a = (a_1, \ldots, a_n) \quad (13.5)$$

such that the following conditions are satisfied:

$$H\left(q, \frac{\partial S}{\partial q}(q; a)\right) = 0, \quad (13.6)$$

and

$$\det\left(\frac{\partial S}{\partial a^i \, \partial a^j}\right) \neq 0. \quad (13.7)$$

Let:

$$M \text{ be the space of variables } (q, p), \quad (13.8)$$

and:

$$\omega = dp_1 \wedge dq^1 + \ldots + dp_n \wedge dq^n. \quad (13.9)$$

Let **C** be the exterior differential system (13.10)
(i.e., the ideal in the Grassmann algebra **D**(M)
of differential forms on M, which is closed under
exterior differentiation d) generated by the
2-form ω and the 0-form H.

Geometric PDE Theory

Thus, for each:

$$a \in R^n, \tag{13.11}$$

the map:

$$\phi_a: q \to (q, p, \frac{\partial S}{\partial q}) \tag{13.12}$$

is a submanifold map of:

$$Q \to M. \tag{13.13}$$

It defines an n-dimensional integral submanifold of C, i.e.,

$$\phi_a^*(C) = 0. \tag{13.14}$$

As a varies, we obtain:

a foliation of M by n-dimensional submanifolds. (13.15)

since ϕ_a is a submanifold map [a fact guaranteed by condition (13.7)].

Remark. In the contemporary literature on symplectic manifolds, this is often called a **Lagrangian foliation** (with historical justification, since it was Lagrange who introduced the concept of "complete solution" for first-order nonlinear partial differential equations).

To introduce the notion of "singular solution" of a first-order partial differential equation" (as developed by Goursat [6] and Forsythe [5]), suppose that (13.7) is only satisfied in certain regions, but not identically.

Definition. The solutions of (13.6) are of the following form:

$$q \to S(q; a(q)) \tag{13.16}$$

such that

$$\det\left(\frac{\partial^2 S}{\partial a^i \, \partial a^j}(q; a(q))\right) = 0 \tag{13.17}$$

for all q,

are called **singular solutions** of the PDE 13.1. In this way of regarding them, they are dependent on the preliminary choice of "complete solution."

E. Vessiot, in 1924, [24] proposed a generalization of this classical concept of "complete solution." I will now rephrase Vessiot's idea in terms of the Ehresmann jet spaces.

Definition: Let:

$$\sigma: J^r(X, Y) \to Z \tag{13.18}$$

be the symbol of an rth-order differential operator with domain manifold X and range manifold Y. Set:

$$DE(\sigma) = \sigma^{-1}(0) \tag{13.19}$$

be the subset of the jet space, which defines the "differential equation." Let:

$$\gamma: J^r(X, Y) \to W \tag{13.20}$$

be a submersion mapping of $J^r(X, Y)$ onto a manifold W.

Remark. Recall that 'submersion' means that the induced linear mapping on tangent bundles is **onto**.

For each $w \in W$, the fiber:

$$\gamma^{-1}(w) \tag{13.21}$$

Geometric PDE Theory

is then a submanifold of $J^r(X, Y)$.

Definition. γ is said to define a **Lagrange-Vessiot fibration** of $DE(\sigma)$ if the following conditions are satisfied:

(a) dimension $\gamma^{-1}(w) = \dim X$ (13.22)
   for all $w \in W$,

(b) $\gamma^{-1}(w) \subset DE(\sigma)$, (13.23)
   for all $w \in W$,

(c) For each $w \in W$, the Source projection map: $\gamma^{-1}(w) \to X$ is a local diffeomorphism, (13.24)

(d) For eachl $w \in W$, the contact forms on $J^r(X, Z)$ vanish when restricted to $\gamma^{-1}(w)$. (13.25)

In his fundamental paper [24] Vessiot gives sufficient conditions for the local existence of such complete solutions that we will not go into it at this point. [The conditions are really that the exterior differential system generated by the contact form, restricted to $DE(\sigma)$, assumed to be a submanifold, are in **involution** in Cartan's sense. [1, 2, 25]

I will now assume that Lagrange-Vessiot fibrations exist, and examine the consequences. I will be particularly concerned with conditions that **one** such fibration determines the "general solution" of $DE(\sigma)$. This is what happens in the case of a nonlinear first-order partial differential equation: Such a description is called the Method of Lagrange and Charpit [6, 5] in the classical literature.

14. GENERAL SOLUTIONS THAT ARE GENERATED FROM ONE PARTICULAR COMPLETE SOLUTION. LAGRANGE-VESSIOT FIBRATION BY THE METHOD OF LAGRANGE AND CHARPIT

Continue with the notation of Section 13. Let:

$$\gamma: J^r(X, Y) \to W \qquad (14.1)$$

be a submersion map

that defines a Lagrange-Vessiot fibration [i.e., satisfies conditions (13.22) — (13.25)] for the partial differential equation associated with the symbol map:

$$\sigma: J^r(X, Y) \to Z. \qquad (14.2)$$

Let:

$$U \text{ be an open subset of } X, \qquad (14.3)$$

and let:

$$\phi: U \to Y \qquad (14.4)$$

be a map which is a solution of the partial differential equation generated by σ, i.e. the following condition is satisfied:

$$j^r(\phi)(U) \subset DE(\sigma). \qquad (14.5)$$

We can then construct the map

$$\gamma j^r(\phi): U \to W. \qquad (14.6)$$

Thus, to the solution ϕ we assign the following subset:

$$\gamma j^r(\phi)(U) \subset W \qquad (14.7)$$

For certain ϕ's, this will be a **submanifold** of Z. The classical method of Lagrange and Charpit attempts to reverse this process, i.e.

Geometric PDE Theory

Assigning a solution ϕ to the partial differential equation to certain submanifolds of W. (14.8)

In certain favorable cases, one can generate a general solution to the partial differential equation in this way.

In order to understand better when the situation is "favorable" to the Lagrange-Charpit method, it is most convenient to introduce more explicitly the geometric ideas of Cartan's theory of exterior differential systems.

15. THE METHOD OF LAGRANGE-CHARPIT IN THE CONTEXT OF CARTAN'S THEORY OF EXTERIOR DIFFERENTIAL SYSTEMS

Let X be a manifold with a set **I** of differential forms on X which define an **exterior differential system**, i.e. the following condiions are satisfied:

I is an ideal of algebraic sense for the Grassmann algebra structure of differential forms on X, (15.1)

plus:

I is closed under the exterior derivative operation d. (15.2)

For $x \in X$, let:

X_x be the **tangent space to X at x**; (15.3)

X_x^d is its dual space, the **space of covectors at x**. (15.4)

Let :

$E(X_x^d)$ be the exterior algebra of the vector space X_x^d. (15.5)

I(x) be the set of values at x of the forms ion **I**. (15.6)

$I(x)$ is then an ideal in $E(M_p^d)$. (15.7)

Recall that a submanifold map:

$$\phi: Y \to X \qquad (15.8)$$

is said to be an **integral submanifold** of I if:

$$\phi^d(I) = 0. \qquad (15.9)$$

A linear subspace α of the tangent space X_x is said to be an **integral element** of **C** if:

$$\theta(\alpha) = 0. \qquad (15.10)$$
$$\text{for all } \theta \in I$$

Hence, the condition (15.9) that the map $\phi: Y \to X$ be an integral submanifold is equivalent to the following:

$$\phi_*(Y_y) \text{ is an integral element,} \qquad (15.11)$$
$$\text{for each } y \in Y.$$

Definition. α is said to be a **maximal integral element** if it is contained in no larger integral element.

Definitions: A tangent vector $v \in X_x$ is **Cauchy characteristic** for the ideal **I** if the following condition is satisfied:

$$v \rfloor I \subset I(x). \qquad (15.12)$$

$C(I)$ denotes the set of such Cauchy characteristic tangent vectors. It defines a linear subbundle:

Geometric PDE Theory

$$x \to \mathbf{C}(I)(x) \tag{15.13}$$

of the tangent vector bundle T(X) (but one which may have varying dimensions of the fibers, i.e., some sort of "singular" vector bundle).

A vector field $V \in \mathbf{U}(X)$ is said to be **Cauchy characteristic** if the values V(x), for each x, are Cauchy characteristic.

Geometric PDE Theory

Theorem 15.1: If $\alpha \subset X_x$ is a maximal integral element at x of the exterior system I, then:

$$C(I)(x) \subset \alpha. \tag{15.14}$$

Proof: Let $v \in X_x$ be Cauchy characteristic for X, i.e., (15.12) is satisfied. Then, for:

$$v_1, \ldots, v_m \in \alpha, \theta \in I^{m+1},$$

we have:

$$\theta(v, v_1, \ldots, v_m) = (v \,\rfloor\, \theta)(v_1, \ldots, v_m)$$
$$= 0, \tag{15.15}$$

using (13.2) and the hypothesis that α is an integral element. This shows that

$$\alpha + (v) \tag{15.16}$$

is also an integral element. By hypothesis, α is maximal, hence:

$$\alpha + (v) \subset \alpha \tag{15.17}$$

whence:

$$v \in \alpha. \tag{15.18}$$

q.e.d.

Geometric PDE Theory

Definition: A submersion map $\gamma: X \to Z$, with Z a manifold, is said to be a **Lagrange-Vessiot Submersion** (relative to the exterior system **I** on M) if the following conditions are satisfied:

(a) For each $a \in A$, the fiber $\gamma^{-1}(a)$ (15.19)
is an integral submanifold of **I**.

(b) For each $x \in X$, the tangent subspace $\gamma_*^{-1}(0)$ is (15.20)
a **maximal** integral element of **I**.

Theorem 15.2: Suppose that:

$$\gamma: X \to Z \qquad (15.21)$$

is a Lagrange-Vessiot submersion, in the sense of the above definition. Then, the Cauchy characteristic tangent vectors and vector fields are tangent to the fibers of γ.

Proof: Follows from Theorem 15.1.

q.e.d.

Now, let:

$$\gamma: X \to Z$$

again be a fixed Lagrange-Vessiot submersion and let:

$$\phi: Y \to X \qquad (15.22)$$

be an integral submanifold of **C** such that the following conditions are satsified:

$\dim Y = \dim$ fibers of γ;

For each $y \in Y$, (15.23)
the integral element $\phi_*(Y_y)$ is maximal.

Consider the composed map:

$$\gamma\phi: Y \to Z. \tag{15.24}$$

Theorem 15.3: For each $y \in Y$, the Cauchy characteristic vectors $C(I)\,(\phi(y))$ are tangent to the submanifold $\phi(Y)$ of X, and are mapped onto zero under γ_*.

Proof: This follows from Theorem 15.1, i.e., that the Cauchy characteristic vectors are contained in **every** maximal integral element.

q.e.d.

Definition: The integral submanifold:

$$\phi: Y \to X \tag{15.25}$$

is said to be of **Lagrange-Charpit** type relative to the Vessiot submersion:

$$\gamma: X \to Z \tag{15.26}$$

if, for each $y \in Y$:

$$\phi_*(Y_y) \text{ is a maximal integral element of } I, \tag{15.27}$$

and if:

$$\text{the dimension of the kernel } (\gamma\phi)_*^{-1}(0) \tag{15.28}$$
of the map $(\gamma\phi): Y \to Z$ is equal to the
dimension of the Cauchy characteristic vectors of I.

Theorem 15.4: If the following condition is satisfied:

$$\text{the dimension of the Cauchy characteristic vectors} \tag{15.29}$$
of I is constant over X,

and if:
$$\phi: Y \to X \qquad (15.30)$$
is an integral submanifold of Lagrange-Charpit type,

then:

$\gamma\phi$ is a submersion mapping from Y to a submanifold of Z. (15.31)

The dimension of this submanifold of Z is (15.32)
equal to the dimension of P minus the
dimension of the Cauchy characteristic vectors.

Proof: Again, this follows from our hypotheses and the implicit function theorem, since the map $\gamma\phi: Y \to Z$ is of constant rank.

q.e.d.

Thus, we have presented the Lagrange-Charpit method in modern language. In favorable cases, it can be inverted, assigning to each submanifold of certain dimension on Z an integral manifold of **I**. In certain cases described in the classical literature, this will give a "general solution" in the Ampère sense described above. However, there are many partial differential equations, for which the construction of the "general solution" is more complicated geometrically.

16. THE CLASSICAL TECHNIQUES OF THE "INTERMEDIATE INTEGRAL" FOR MONGE-AMPÈRE SECOND-ORDER PARTIAL DIFFERENTIAL EQUATIONS

In the 19th century, there was extensive work on finding conditions that given second-order partial differential equations could exhibit families of solutions that were also solutions of a first-order partial differential equation. This was known as the **method of the intermediate integral**, and is discussed by Goursat [6] and Forsythe [5]. In this section, I will investigate in modern terms certain aspects of this method, elaborating on a brief discussion in [9].

Geometric PDE Theory

Let X and Y be manifolds.

Definition: Let:

$$\sigma: J^2(X, Y) \to Z$$

be the symbol of a second-order differential operator. A first-order, scalar-valued symbol:

$$f: J^1(X, Y) \to R$$

is said to be an **intermediate integral** of the system defined by σ if every solution $\phi: X \to Y$ of the differential equation:

$$j^1(\phi)d(f) = 0 \qquad (16.1)$$

also satisfies:

$$j^2(\phi)d(\sigma) = 0. \qquad (16.2)$$

There is an obvious practical interest in finding such "intermediate integrals." For example, physical systems often lead to second-order partial differential equations. Solutions of (16.1) can be found using ordinary differential equations. Thus, the method, if it works, will generate families of solutions of the physical equations which may be obtained by solution of ordinary differential equations.

Geometric PDE Theory

An example of such a situation is provided by factoring of second-order operators. Suppose:

$$Y = R \tag{16.3}$$

and that:

$$\Delta: \mathbf{F}(Y) \to \mathbf{F}(Y) \tag{16.4}$$

is a second-order differential operator such that:

$$\Delta(f) = \sigma j^2 (f). \tag{16.5}$$

Suppose that:

$$\Delta = \Delta_2 \Delta_1, \tag{16.6}$$

where Δ_2, Δ_1 are finite-order operators.

Let:

$$f: J^1(X, R) \to R \tag{16.7}$$

be the symbol of Δ_1. Then, f is an "intermediate integral" of σ.

In this section, we shall pursue the "Monge-Ampère" method [2, 3, 5, 6, 24] of constructing second-order equations which admit such "intermediate integrals."

Geometric PDE Theory

Let:

$$\mathbf{C} \subset \mathbf{D}(J^1(X, Y)) \tag{16.8}$$

be the Grassmann algebra ideal generated by:

$$\text{the contact forms on } J^1(X, Y). \tag{16.9}$$

Suppose:

$$\dim Y = m, \tag{16.10}$$

$$\dim X = n. \tag{16.11}$$

Definition: The second-order differential equation defined by the symbol map:

$$\sigma: J^2(X, Y) \to Z \tag{16.12}$$

is of **Monge-Ampère type** if:

there is a 2-form ω on $J^1(X, Y)$ such that the following conditions are satisfied:

$$\omega^n \wedge \theta^1 \wedge \ldots \wedge \theta^m = 0 \text{ for all } \theta^1, \ldots, \theta^m \in C. \tag{16.13}$$

($\omega^n = \omega \wedge \ldots \wedge \omega$ denotes the exterior product of n copies of ω).

Each map $\phi: X \to Y$ such that
$$j^1(\phi^*)(\omega) = 0 \tag{16.14}$$
also satisfies

$$j^2(\phi)^*(\sigma) = 0.$$

We will only deal here with the case:

Geometric PDE Theory

$$\dim X = n = 2,$$

$$\dim Y = m = 1, \qquad (16.15)$$

i.e., we consider:

> partial differential equations in two independent variables and one dependent variable.

Let:

> J be the Grassmann algebra ideal of forms on $J^1(X, Y)$ generated by C and ω. $\qquad (16.16)$

For $x \in J^1(X, Y)$, let:

> $E(J)(z)$ be the characteristic tangent vectors of this Grassmann ideal J, i.e., $\qquad (16.17)$

$$E(J)(z) = \{v \in J^1(X, Y)_z : v \rfloor J \subset J\}. \qquad (16.18)$$

Let us make the following regularity assumption:

> $\dim E(J)(z)$ is constant as z ranges over $J^1(X, Y)$. $\qquad (16.19)$

(Note that J is not closed under d, so that it is not an "exterior differential system." Of course, it can be closed up to generate such a system.)

We see immediately that the following relations are satisfied:

> The dimension of $J^1(X, Y)$ is five. $\qquad (16.20)$

> C is generated by a single 1-form. $\qquad (16.21)$

Definition. Let:

$$E(\mathbf{C}) \subset T(J^1(X, y)) \tag{16.22}$$

be the submodule of the tangent bundle to $J^1(K, Y)$ defined by the annihilation of the forms in **C**.

Remark. Since **C** is generated by 1-forms, $E(\mathbf{C})$ can be defined as the Cauchy characteristic vectors of **C**, which again is a Grassmann ideal, but not closed under d.

We see that:

The fibers of $E(C)$ are then **four**-dimensional. (16.23)

Relation (16.23) then means that:

ω, restricted to $E(C)$, is a skew-symmetric form of rank 2. (16.24)

Thus it can be written locally as follows:

$$\omega = \theta_1 \wedge \theta_2 + \theta_3 \wedge \theta, \tag{16.25}$$

where:

θ is a 1-form generating **C**. (16.26)

The vectors in $E(J)$ are then defined by the following Pfaffian equations:

$$\theta_1 = \theta_2 = \theta = 0. \tag{16.27}$$

To avoid degenerate cases, let us assume that

$$\omega \notin \mathbf{C}. \tag{16.28}$$

Geometric PDE Theory

Then the forms (16.27) are linearly independent, i.e.,

$$\dim E(J) = 2. \tag{16.29}$$

Definition: Let $\phi: X \to Y$ be a map which is a solution of the differential equation:

$$j^1(\phi)d(\omega) = 0. \tag{16.30}$$

The curves:

$$t \to x(t) \tag{16.31}$$

on X such that

$$j^1(\phi)d(\theta_1)\left(\frac{dx}{dt}\right) = 0 \tag{16.32}$$

or

$$j^1(\phi)d(\theta_2)\left(\frac{dx}{dt}\right) = 0 \tag{16.33}$$

are the **characteristics**.

Let:

$$(x^i, y, y_i), i = 1, 2, \tag{16.34}$$

be the natural jet coordinates on $J^1(X, Z)$ with respect to coordinates (x^i) for X and a coordinate y for Y. Then:

$$\theta = dy - \Sigma_i y_i \, dx^i. \tag{16.35}$$

Thus, mod θ, θ_1, and θ_2 can be written as follows:

Geometric PDE Theory

$$\theta_1 = \Sigma_i A^i \, dy_i + A_i \, dx^i, \tag{16.36}$$

$$\theta_2 = \Sigma_i B^i \, dy_i + \Sigma_i B_i \, dx^i, \tag{16.37}$$

where the A's and B's are functions of (s^i, y, y_i). Hence,

$$\begin{aligned}\theta_1 \wedge \theta_2 &= (A^1 B^2 - A^2 B^1) \, dy_1 \wedge y_2 \\ &\quad + (A^i B_j - A^j B_i) \, dy_i \wedge dx^j \\ &\quad + (A_1 B_2 - A_2 B_1) \, dx^1 \wedge dx^2 \\ &\equiv a \, dy_1 \wedge dy_2 + a^i_j \, dy_i \wedge dx^j + \beta \, dx^1 \wedge dx, \end{aligned} \tag{16.38}$$

$$j^1(\phi)d(y) = y(x^1, x^2), \tag{16.39}$$

$$j^1(\phi)d(y_i) = \frac{\partial y}{\partial x^i} \tag{16.40}$$

The differential equation (16.12), in these coordinates, takes on the Monge-Ampère form:

$$a\left(x, y, \frac{\partial y}{\partial x}\right)\left(\frac{\partial^2 y}{\partial x^1 \partial x^1} \frac{\partial^2 y}{\partial x^2 \partial x^2} - \left(\frac{\partial^2 y}{\partial x^1 \partial x^2}\right)^2\right) \\ + F\left(x, y, \frac{\partial^2 y}{\partial x^2}\right)\frac{\partial y}{\partial x} = 0. \tag{16.41}$$

Geometric PDE Theory

Theorem 16.1: Let f: $J^1(X, Y) \to R$ be a function such that

$$df(E(J)) = 0, \, df \notin C. \tag{16.42}$$

Then, any map:

$$\phi: X \to Y \tag{16.43}$$

satisfying the first-order differential equation

$$j^1(\phi)d(f) = \text{const} \tag{16.44}$$

also satisfies the Monge-Ampère equation (16.41), i.e.,

f is an "intermediate integral" of (16.41).

Proof. (16.44) implies that df is a linear combination of θ, θ_1, θ_2, which implies that ω can be written as follows:

$$\omega = df \wedge \theta_1' + \text{(forms in } C\text{)}; \tag{16.46}$$

hence

$$j^1(\phi)^*(\omega) = 0, \tag{16.47}$$

whence, as we have seen, (16.41).

q.e.d.

Geometric PDE Theory

17. THE FAMILIES OF SOLUTIONS OF THE MONGE-AMPÈRE EQUATION GENERATED BY ALL INTERMEDIATE INTEGRALS

I now want to make more precise some remarks on pp. 250-1 of Ref. 9 concerning when the intermediate integrals generate general solutions of Monge-Ampère equations. Again, let X and Y be manifolds with:

$$\dim X = 2, \quad \dim Y = 1.$$

Let:
$$Z = J^1(X, Y). \tag{17.1}$$

Consider:

$$\mathbf{P} \subset \mathbf{D}^1(Z),$$

a nonsingular Pfaffian system on Z, i.e.,

$$\mathbf{P} \text{ is an } \mathbf{F}(Z)\text{-module of 1-differential forms on } Z \tag{17.2}$$

such that the following condition is satisfied:

$$\dim (\mathbf{P}(z)) = \text{const}, \tag{17.3}$$
where z runs over all points of z.

[$\mathbf{P}(z)$ denotes the set of values at z of all forms in $\mathbf{P}$. $\mathbf{P}(z)$ is a linear subspace of Z_z^d, the space of one-covectors to Z at z.]

Set:

$$\mathbf{P}^1 = \{\theta \in \mathbf{P}: d\theta \in \mathbf{D}^1(Z) \wedge \mathbf{P}\}. \tag{17.4}$$

$\mathbf{P}^1$ is another Pfaffian system, (17.5)
called the **first derived system** to $\mathbf{P}$.

Geometric PDE Theory

We can iterate the construction, obtaining a decreasing sequence:

$$\mathbf{P} \supset \mathbf{P}^1 \supset \ldots \tag{17.6}$$

of Pfaffian systems. We will suppose these are **nonsingular** in the sense that:

> Their values at points z are constant, as z ranges over Z. (17.7)

Hence, there is a first integer r such that:

$$\mathbf{P}^r = \mathbf{P}^{r+1} = \mathbf{P}^{r+2} = \ldots . \tag{17.8}$$

Now, suppose that ω is a 2-form on Z such that:

$$\omega \in \mathbf{P} \wedge \mathbf{P}, \tag{17.9}$$

$$\omega \wedge \omega = 0. \tag{17.10}$$

(17.8) and (17.9) guarantee that, locally, ω can be written as:

$$\omega = \theta_1 \wedge \theta_2 \quad \text{with } \theta_1, \theta_2 \in \mathbf{P}. \tag{17.11}$$

We can now define a partial differential equation of Monge-Ampère type for mappings (defined locally) $\phi: X \to Y$ such that:

$$j^1(\phi)^*(\omega) = 0, \tag{17.12}$$

i.e., the maps ϕ satisfying (17.12) are the **solutions**.

Linear algebra, plus (17.10), implies the following result:

Geometric PDE Theory

Theorem 17.1: Suppose ϕ is a map: $X \to Y$ satisfying (17.10). Then,

$$\dim (j^1(\phi)^* (P))(x) = 0 \text{ or } 1. \qquad (17.13)$$
$$\text{for } x \in X$$

Let us now consider:

$$\text{points } x \in X \text{ where the dimension in (17.13) is one.} \qquad (17.14)$$

At points satisfying 17.14, (which form an open subset of X, whose complement is a union of varieties of dimension 0 or 1, if the data is real analytic), the integral curves of $j^1(\phi)^*(P)$ are the **characteristics**.

Remark. In the Monge-Ampère case, the two-parameter family of characteristics for general second-order partial differential equations "degenerate" to a one-parameter family.

Proof. Left to the reader.

Definition: A function $f: Z \to R$ such that

$$df \in \mathbf{P}, \qquad (17.15)$$

is called an **intermediate integral** for the Monge-Ampère equation (17.12).

Geometric PDE Theory

Theorem 17.2: If $f \in \mathbf{F}(Z)$ is an intermediate integral for the Monge-Ampère equation (17.12), then

$$df \in \mathbf{P}^r. \tag{17.16}$$

By (17.16):

$\mathbf{P}^r$ is completely integrable in the Frobenius sense (17.17)
and hence defines a foliation on Z.

f is constant on the leaves of this foliation. (17.18)

Proof: Again, this follows from (17.3) and the fact that:

$$d(df) = 0 \text{ for all } f \in \mathbf{F}(Z).$$

q.e.d.

One can analyze how many intermediate integrals there are, locally, for the Monge-Ampère equation (17.12). Let us suppose that:

$\dim \mathbf{P}^2(z)$ is constant as z varies over Z. (17.19)

$\mathbf{P}^2$ then defines a foliation on Z. (17.20)

Let us suppose that this foliation is **regular** in the sense that:

It has a quotient space $\overline{Z}$, which is a manifold. (17.21)

Thus, we can parametrize the f's satisfying (17.14) by elements of $\mathbf{F}(\overline{Z})$.

To such an f satisfying (17.14), we consider:

$$f: J^1(X, Y) \equiv Z \to R \tag{17.20}$$

as a scalar-valued symbol of a **first-order** partial differential operator.

The solutions of the corresponding first-order partial differential equation will then also be solutions of the Monge-Ampère equation. A general solution for this equation can be found by the Lagrange-Charpit method described earlier; they will be parametrized by curves in the quotient space $\bar{\bar{Z}}$ of a Lagrange-Vessiot submersion. Two f's can also define the submanifold

$$\bar{f} = 0 \tag{17.21}$$

on $\bar{Z}$.

All together we have defined:

$$\text{a manifold } W = \bar{Z} \times \bar{\bar{Z}} \tag{17.22}$$

such that:

the families of solutions of the Monge-Ampère equation are parametrized by **submanifolds** of a certain dimension of W. (17.23)

In case $\mathbf{P} = \mathbf{P}^1$, i.e., the characteristics are Frobenius integrable, this family of solutions will be a **general solution**, in the Ampère sense. (17.24)

Much of the 19th century geometric theory of partial differential equations is motivated by the goal of generalizing this Monge-Ampère method. From the modern point of view, it is important since:

(a) it is "algorithmic," (17.25)

(b) it carries through in the C^∞ case. (17.26)

Remark. Again, it wil be a challenge to extend these ideas to the C-O-R Generalized Function context.

This manifold W also carries a geometric structure, which defines various groups. These groups act then on solutions of the Monge-Ampère equations in ways that go "beyond" the action of groups on the jet spaces associated with X and Y.

Bibliography

1. R. Bryant, S. Chern, R. Gardner, H. Goldschmidt and P. Griffiths, **Exterior Differential Systems,** Springer-Verlag, 1991

2. E. Cartan, **Les systèmes éxterieures et leurs applications géométriques** (Herman, Paris, 1946).

3. G. Darboux, **Lecons sur la théorie générale des surfaces** (Gauthier-Villars, Paris, 1894, 1915).

4. C. Ehresmann: (a) "Sur les structures infinitésimales régularières," Congrés Int. Math. Amsterdam 1, 479-80 (1954); (b) "Connexions infinitésimales," Colloq. Top. Alg. Bruxelles, 29-55 (1950); (c) "Structures infinitésimales et pseudogroupes de Lie," Colloq. Int. C.N.R.S. Géom. Diff. Strasbourg, 97-110 (1953); (d) C. R. Acad. Sci. Paris **240, 246,** 360 (1958) (1954); **241,** 397, 1755 (1955); (e) "Connexions d'ordre supérieur," **Atti 5 Cong. dell'Unione Mat. Italiana, 1955** (Cremonese, Rome, 1956), pp. 326-8; (f) "Categories topologiques et catégories différentiables," Colloq. Géom. Diff. Globale Bruxelles, C.B.R.M., 137-150 (1958); (g) "Groupoides differenciales," Rev. Un. Mat. Argentina Buenos-Aires **XIX,** 48 (1960).

5. A. Forsythe, **Theory of Differential Equations** (Dover, New York, 1959).

Geometric PDE Theory

5a. H. Goldschmidt, (a) "Nonlinear partial differential equations," J. Diff. Geom. **1**, 269-307 (1967); (b) "Sur la structure des équations de Lie," J. Diff. Geom. **6**, 357-73 (1972); **7**, 67-95 (1972).

5b. M. Golubitsky and V. Guillemin, **Stable Mappings and Their Singularities** (Springer-Verlag, New York, 1973).

6. E. Goursat, **Lecons sur l'intégration des équations aux derivés partielles du second ordre** (Gauthier-Villars, Paris, 1890), Vols. I and II.

6a. R. Hermann, Geometric Construction and Properties of Some families of Solutions of Nonlinear Partial Differential Equations, **J. Math. Phys.**, 24, 1983, 510-517.

6b. R. Hermann, "Some differential geometric aspects of the Lagrange variational problem," Ill. J. Math. **6**, 634-73 (1962).

6c. R. Hermann, "The second variation for variational problems in canonical form," Bull. Am. Math. Soc. **71**, 145-8 (1965).

7. R. Hermann, "The second variation for minimal submanifolds," J. Math. Mech. **16**, 473-92 (1966).

8. R. Hermann, **Differential Geometry and the Calculus of Variations** (Math. Sci. Press, Brookline, MA, 1977), 2nd ed.

9. R. Hermann, **Geometry Physics and Systems** (Marcel Dekker, New York, 1973).

10. R. Hermann, **Geometry of Non-Linear Differential Equations, Bäcklund Transformations and Solitons, Part A**, Vol. XII of Interdisciplinary Mathematics (Math. Sci. Press, Brookline, MA, 1976).

11. R. Hermann, "The differential geometry of general mechanical systems from the Lagrangian point of view," J. Math. Phys. **23**, 2077 (1982).

12. R. Hermann, "E. Cartan's geometric theory of partial differential equations," Adv. in Math. **1**, 265-317 (1965).

16. E. Ince, **Ordinary Differential Equations** (Dover, New York, 1956).

16a. A. Kumpera and D. C. Spencer, **Lie Equations**, Ann. of Math. Studies **73** (Princeton U. P., Princeton, NJ, 1972).

13. B. A. Kupershmidt, in **Lecture Notes in Mathematics 775** (Springer-Verlag, New York, 1980), pp. 162-217.

16. P.Libermann and C. M. Marle, **Symplectic Geometry and Analytical Mechanics**, Kluwer, Dordrecht, 1987

17. K. Mackenzie, **Lie Groupoids and Lie Algebroids in Differential Geometry**, Cambridge University Press, 1987

18. E. Picard, **Traité d'analyse** (Gauthier-Villars, Paris, 1922).

19. C. Riquier, **Les systémes d'equations aux derivées partielles** (Gauthier-Villars, Paris, 1910).

20. D. J. Saunders, **The Geometry of Jet Bundles,** Cambridge University Press, 1989

21. D. C. Spencer, "Overdetermined systems of linear partial differential equations," Bull. Am. Math. Soc. **75**, 179-239 (1969).

22. E. Vessiot, "Sur une théorie nouvelle des problèmes généraux d'integration," Bull. Soc. Math. Fr. **52**, 336-96 (1924).

23. A. M. Vinogradov, "Many-valued solutions, and a principle for the classification of nonlinear differential equations," Dokl. Akad. Nauk SSSR **210**, 11-14 (1973) [Sov. Math. Dokl. **14**, 661-5 (1973)], MR 50 No. 12916.

24. E. von Weber, G. Floquet, and E. Goursat, "Proprietés générales des systems d'equations aux derivées partielle," in **Encyclopédie des Sciences Mathématiques** (Gauthier-Villars, Paris, 1913), Tome II, Vol. 16.

25. K. Yang, **Exterior Differential Systems and Equivalence Problems,** Kluwer, 1992.

CHAPTER 4

THE POISSON-CARTAN ALGEBRA STRUCTURE AND PROLONGATION CONCEPTS ASSOCIATED WITH A SMOOTH, CLOSED 2-FORM ON A SMOOTH MANIFOLD

From the point of view I have utilized in my work since my Lincoln Lab days of 1959-61, Mechanics (and by this I mean much of field-theoretic physics as well!) involves two types of Geometric Structures: various Lie -Cartan - Ehresmann- Spencer pseudogroups and Cartan-Ehresmann Connections. In this Chapter, I will adopt a general point of view and study the Geometric Structures associated with a closed 2-form on a manifold. The Bibliography on Mechanics at the end of this Chapter will serve for the rest of this Volume.

1. Introduction.

In this Chapter, I study some algebraic and geometric properties of the Geometric Structure defined by a closed 2-form which may have non-zero characteristic vectors. I will be concerned with three main questions:

a). The Algebraic Structure defining a Generalized Poisson Bracket. This work will be continued in a later Chapter dealing with Cartan 'constrained Poisson Bracket'.

b). The concept of 'prolongation' associated with this type of Geometric Structure. (This is also sometimes called 'reduction' in recent Applied literature. However, Lie, Cartan and Vessiot used the more general and consistent term 'prolongation')

c). The role that the Cauchy Characteristic vectors and orbit curves play in the study of Mechanical Systems.

My immediate Applied aims are two fold:

Poisson-Cartan

Make precise the pseudogroup structure involved, in preparation for Numerical work. For example, the 'philosophy' of choosing 'Sympletic Integrators' is that the Numerical Approximations should satisfy the same constraints as the ODE's which are being approximated. One such 'constraint' is that of belong to the Lie Pseudogroup associated with the given ODE system.

Interpret the Regularization methods of Levi-Civita, Kustaamheimo and Stiefel in general 'prolongation' terms. I will begin this work in a later Chapter.

I will suppose given the following data:

$$\text{A smooth manifold X} \tag{1.1}$$

A smooth differential 2-form ω on X such that: $\tag{1.2}$

$$d\omega = 0$$

$$\mathbf{F}(X) = \text{smooth real-valued functions on X} \tag{1.3}$$

$\mathbf{U}(X)$ = smooth vector fields on X, considered as an $\mathbf{F}(X)$- module and as a real Lie algebra under Jacobi-Lie bracket [,]. $\tag{1.4}$

$$\mathbf{D}^r(X) = \text{smooth differential forms of degree r.} \tag{1.5}$$

$\mathbf{P}(\omega)$ = set of pairs (V, f) consisting of:

$$\text{A smooth vector field V on X} \tag{1.6}$$

A smooth real-valued function f on X such that:

$$df = V \lrcorner \, \omega \tag{1.7}$$

Define an algebra structure over the real number field R on $\mathbf{P}(\omega)$:

$$\{\,,\}: \mathbf{P}(\omega) \times \mathbf{P}(\omega) \longrightarrow \mathbf{P}(\omega) \tag{1.8}$$

by means of the following formula:

$$\{(V, f), (V', f')\} = \{([V, V'], 1/2[V(f') - V'(f)]\} \tag{1.9}$$

Definition. $\mathbf{P}(\omega)$, with the algebra structure defined by 1.8-1.9, is called the **Poisson-Cartan algebra structure associated with the closed 2-form** ω.

Remark. I have given this algebra structure the name 'Poisson-Cartan' because:

> On the one hand, it is isomorphic to the usual Poisson Bracket structure in case ω has no non-zero Cauchy characteristic vectors, i.e. defines a **symplectic structure** on X, in the usual sense.
>
> It captures certain features described in his unique way by Elie Cartan in his 1921 treatise "Lecons sur les Invariants Integraux". Various aspects of this Structure can also be found in my work, beginning with my book "Lie Algebras and Quantum Mechanics" published in 1970.

2. **Some properties of the Poisson-Cartan algebra**

Associated with the 2-form ω satisfying 1.2, we have the pseudogroup of its local automorphisms. The Lie algebra of this pseudogroup is a Lie algebra of vector fields on X, which we define as follows:

$$\mathbf{S}(\omega) = \{V \in \mathbf{U}(X): V(\omega) = 0\} \tag{2.1}$$

Poisson-Cartan

Theorem 2.1 $S(\omega)$ is a Lie sublgebra of the Lie algebra $U(X)$ (under Jacobi-Lie Bracket) of all smooth vctor fields on X.

Proof. This follows from the following identity between iterated Lie derivative and Jacobi-Lie Bracket:

$$[V, V'](\omega) = V(V'(\omega)) - V'(V(\omega)) \tag{2.2}$$

Theorem 2.2. Define a map π from the Poisson-Cartan algebra $P(\omega)$ to $U(X)$ as follows:

$$\pi(V, f) = V. \tag{2.3}$$

Then, π is an algebra homomorphism.

Proof. Follows from 1.9.

Theorem 2.3. For:

$$(V, f), (V', f') \; \varepsilon \; P(\omega), \tag{2.4}$$

we have:
$$V(f') = - V'(f) \tag{2.5}$$

Hence, the Poisson-Cartan Bracket $\{\,,\,\}$ formula1.9 takes the following form:

$$\{(V, f), (V', f')\} = \{([V, V'], V(f')\} \tag{2.6}$$

Proof. With 2.4 satisfied, we have:

$$V(f') = V \rfloor df' =, \text{ using } 1.7, V \rfloor V' \rfloor \omega = \omega(V', V) \tag{2.7}$$

Since ω is a 2-differential form, we have:

$$\omega(V', V) = - \omega(V, V') \tag{2.8}$$

2.7 and 2.8 prove 2.5.

q.e.d.

Theorem 2.4. The algebra homomorphism π defined by 2.3 satisfies the following relation:

$$\pi(\mathbf{P}(\omega)) \subset \mathbf{S}(\omega) \tag{2.9}$$

Proof. Let $(V, f) \; \varepsilon \; \mathbf{P}(\omega)$, i.e.

$$df = V \rfloor \omega. \tag{2.10}$$

Then, using the Cartan Family Identity,

$$V(\omega) = V \rfloor d\omega + d(V \rfloor \omega)$$

$$= V \rfloor 0 + d(df) = 0, \text{ i.e:}$$

$$\omega \; \varepsilon \; \mathbf{S}(\omega).$$

q.e.d

3. The traditional Hamiltonian Mechanics in terms of Poisson-Cartan structures.

Let:

$$p, q \text{ denote points of } R^n. \tag{3.1}$$

Classically, the ODE's of Mechanics can (at least for Lagrangian systems) be written in Hamiltonian form:

$$dq/dt = H_p$$
$$dp/dt = -H_q, \tag{3.2}$$

where:

$$H: (p, q, t) \longrightarrow H(p, q, t) \tag{3.3}$$

is a map from $R^n \times R^n \times R \longrightarrow R$, with:

$$\begin{aligned} n &= \text{'number of degrees of freedom'} \\ &= \text{dimension of the 'configuration space'} \end{aligned} \tag{3.4}$$

Remark. In 3.2, subscripts denote 'partial derivatives'.

The ODE system 3.2 is to be solved for a curve

$$\{t \longrightarrow (p(t), q(t))\} \tag{3.5}$$

in R^{2n}.

In order to rewrite the ODE system 3.2 in terms of differential forms, set:

$$X = R^{2n+1} = \{(p, q, t): p, q \in R^n, t \in R\} \tag{3.6}$$

Introduce the following indices with the indicated range:

$$1 \leq i, j, k, \ldots \leq n \tag{3.7}$$

Define elements '$\mathbf{p}^i$', '$\mathbf{q}^i$', '$\mathbf{t}$' of $F(X)$ as follows:

$$\mathbf{p}^i(p, q, t) = \text{i-th component of the n-vector 'p'} \tag{3.8}$$

$$\mathbf{q}^i(p, q, t) = \text{i-th component of the n-vector 'q'} \tag{3.9}$$

$$\mathbf{t}(p, q, t) = t \tag{3.10}$$

Poisson-Cartan

Theorem 3.1. Let θ be the following 1-form on R^{2n+1}:

$$\theta = \Sigma_i p^i dq^i - H dt \tag{3.11}$$

Then:

$$d\theta = \Sigma_i dp^i \wedge dq^i - dH \wedge dt \tag{3.12}$$

Also, $d\theta$ can be written in the following form:

$$d\theta = \Sigma_i \alpha^i \wedge d\beta^i \tag{3.13}$$

where:

$$\alpha^i = dp^i - [\partial H/\partial q^i] dt \tag{3.14}$$

$$\beta^i = dq^i + [\partial H/\partial p^i] dt \tag{3.15}$$

Then, a curve $\{\sigma: t \longrightarrow \sigma(t) = (p(t), q(t), t)\}$ is a solution of Hamilton's Equations iff.

$$\sigma^d(\alpha^i) = \sigma^d(\beta^i) = 0, \tag{3.16}$$

i.e.

σ is an integral submanifold of the Exterior Differential System generated by the 1-forms $\{\alpha^i, \beta^i\}$. (3.17)

Proof. Exterior-differentiate both sides of 3.11, obtaining:

$$\begin{aligned} d\theta &= \Sigma_i dp^i \wedge dq^i - dH \wedge dt \\ &= \Sigma_i dp^i \wedge dq^i - \Sigma_i [\partial H/\partial q^i] dq^i \wedge dt - \Sigma_i [\partial H/\partial p^i] dp^i \wedge dt \end{aligned} \tag{3.18}$$

Use 3.14-3.15:

$$\Sigma_i \alpha^i \wedge d\beta^i = \Sigma_i \left(dp^i - [\partial H/\partial q^i] dt \right) \wedge \left(dq^i + [\partial H/\partial p^i] dt \right) \tag{3.19}$$

Multiplying out the right-hand side of 3.19 using the algebraic rules of Exterior Algebra, we see that 3.13 is satisfied.

We also see from 3.14-3.15 that the curve σ satisfies 3.16 if and only if the curve $\{t \rightarrow (p(t), q(t)\}$ satisfies the Hamilton Equations 3.2.

q.e.d.

4. Cauchy characteristic vector fields and curves associated with a closed 2-form.

Cartan himself introduced the concept of 'Cauchy Characteristic' (Vector, Vector Fields, Exterior Differential Systems, Submanifolds, Curves, ...) as a grand generalization of the classical notion of 'characterstics of a PDE system'. He showed that the concept had an especial importance for the Foundations of Mechanics and the Calculus of Variations. The 'classical' foundation for his use of 'characteristics' is contained in the following observation:

> The Hamilton ODE's of Mechanics are the (Cauchy) (4.1)
> Characteristics of the Hamilton-Jacobi First Oder PDE
> associated with the Mechanical System.

What is especially useful about Cartan's formulation is that the 'Cauchy Characteristic' concept is so algebraically-elegant and 'coordinate-free'. I will now review certain properties of the Cauchy Characteristics of the Exterior Differential System associated with a closed 2-form.

Definitions. Let ω be a smooth closed 2-form on a smooth manifold X. A **tangent vector** v to X is **Cauchy Characteristic** iff:

$$v \lrcorner \omega = 0. \qquad (4.2)$$

A **tangent vector field** V to X is **Cauchy Characteristic** iff:

$$V \lrcorner \omega = 0. \qquad (4.3)$$

A **curve** $\{\sigma: [a, b] \to X; t \to \sigma(t)\}$ in X is **Cauchy Characteristic** iff:

$$[d\sigma/dt](\sigma(t)) \,\rfloor\, \omega = 0 \text{ for all } t, \quad (4.4)$$

i.e. iff:

The tangent vectors $\{t \to d\sigma/dt\}$ to the curve σ are all Cauchy Characteristic. $\quad (4.5)$

Theorem 4.1. Let $\{H(p, q, t); p, q \in R^n\}$ be data defining a Mechanical System with 'n-degrees of freedom'. Set:

$$\theta = \Sigma_i p^i dq^i - H dt \quad (4.6)$$

Then:

$$d\theta = \Sigma_i dp^i \wedge dq^i - dH \wedge dt \quad (4.7)$$

A curve of the following form in R^{2n+1}:

$$\{t \to \sigma(t) = (p(t), q(t), t)\} \quad (4.8)$$

is a Cauchy Characteristic Curve for the 2-form '$d\theta$' iff:

The curve $\{t \to p(t), q(t)\}$ in R^{2n} is a solution of the Hamilton ODE system $\quad (4.9)$

Proof. Using 3.13 we have:

$$[d\sigma/dt](\sigma(t)) \,\rfloor\, \omega = [d\sigma/dt](\sigma(t)) \,\rfloor\, \Sigma_i \alpha^i \wedge d\beta^i$$

$$= \Sigma_i \big([d\sigma/dt] \,\rfloor\, \alpha^i\big) \wedge d\beta^i - \Sigma_i \alpha^i \wedge \big([d\sigma/dt] \,\rfloor\, [d\beta^i]\big) \quad (4.10)$$

Thus, if $\{t \to \sigma(t)\}$ is a solution of Hamilton's Equations, then:

$$[d\sigma/dt](\sigma(t)) \,\rfloor\, \omega = 0. \quad (4.11)$$

The steps are reversible, so that if 4.1 is satisfied, then so is 3.2.

q.e.d.

5. The Cauchy Characteristic foliation associated with a closed 2-form.

Suppose again that 1.1 and 1.2 are satisfied, i.e. ω is a smooth, closed 2-form on the smooth manifold X. Set:

$$\text{For } x \in X, \mathbf{C}(\omega, x) = \{v \in X_x : v \rfloor \omega = 0\} \tag{5.1}$$

$$\mathbf{C}(\omega) = \{V \in \mathbf{U}(X) : V \rfloor \omega = 0\} \tag{5.2}$$

The elements of $\mathbf{C}(\omega, x)$ are called the Cauchy Characteristic Vectors at x, and the elements of $\mathbf{C}(\omega)$ are the Cauchy Characteristic Vector Fields.

Theorem 5.1. $\mathbf{C}(\omega)$ is an $F(X)$-submodule of $F(X)$, i.e forms a **vector field system** on X.

Proof. If $V \in \mathbf{C}(\omega)$, i.e. 5.2 is satisfied, and $f \in F(X)$ then:

$$(fV) \rfloor \omega = f(V \rfloor \omega) = 0, \text{ i.e.}$$

$$fV \in \mathbf{C}(\omega).$$

q.e.d

Theorem 5.2. $[\mathbf{C}(\omega), \mathbf{C}(\omega)] \subset \mathbf{C}(\omega)$ \hfill (5.3)

i.e. $\mathbf{C}(\omega)$ defines a Frobenius Integrable Vector Field System on X.

Proof. Let $V, V' \in \mathbf{C}(\omega)$. Then:

$$V'(\omega) = d(V' \rfloor \omega) = d0 = 0. \tag{5.4}$$

Hence:

$$0 = V'(V \lrcorner \omega) = [V',V\,] \lrcorner \,\omega,$$

i.e.:

[V',V] is a Cauchy Characteristic vector field for ω,

and 5.3 is satisfied.

q.e.d.

Definitions. At this point, my Theory of (Possibly) Singular Foliations, briefly developed in [6], can be utilized. The continuous, piecewise-smooth Cauchy Characteristic curves of ω in X, i.e. the curves in X satisfying 4.4-4.5 are then the **orbit** or **integral curves** of the Frobenius Integrable Vector Field System $C(\omega)$. The **Leaves of the Foliation associated with the Frobenius Integrable Vector Field System** $C(\omega)$ are the equivalence classes under the following equivalence relation:

> Two points of X are $C(\omega)$ -**Equivalent** iff. (5.5)
> they can be joined by a continuous, piecewise-smooth
> Cauchy Characteristic curves of ω.

$C(\omega)$ is said to be **non-singular** iff:

$$\dim C(\omega, x) = \text{constant as x ranges over X.} \quad (5.6)$$

A submanifold Y of X is said to be an **integral submanifold** of $C(\omega)$ iff:

For all points $y \in Y$, we have: $Y_y \subset C(\omega, x)$ (5.7)

A submanifold Y of X is said to be a **leaf** of $C(\omega)$ iff:

> Y is a **connected** submanifold of X which is an (5.8)
> integral submanifold of $C(\omega)$ in the sense that 5.7 is
> satisfied, plus the following condition:
>
>> For each pair (y, y') of points of Y, Y contains (5.9)
>> all the continuous, Cauchy Characteristic curves
>> of ω which begin at y and end at y'.

Theorem 5.3. Suppose that one of the following conditions is satisfied:

$C(\omega)$ is non-singular in the sense that 5.6 is satisfied. (5.10)

For each $x \in X$,
$$C(\omega, x) = \{V(x): V \in C(\omega)\}, \quad (5.11)$$
and **either**:
$$(X, \omega) \text{ is real-analytic} \quad (5.12)$$
or
$$C(\omega) \text{ is locally-finitely generated, as defined in [6].} \quad (5.13)$$

For each smooth Cauchy Characteristic curve $\{t \dashrightarrow x(t)\}$ of ω, the following condition is satisfied:

$$\dim C(\omega, x(t)) = \text{constant for all } t. \quad (5.14)$$

Then, through each point of X, there is a leaf of the Cauchy Characteristic system of ω.

Proof. This is a consequence of the results in my Singular Foliations paper [6].

6. Mechanical systems and their symmetries in terms of Poisson-Cartan structures.

Recapitulating, let us suppose given the following data:

A smooth manifold X (6.1)

A smooth differential 2-form ω on X such that:

$$d\omega = 0 \quad (6.2)$$

A pair (V_H, H) consisting of:

A smooth vector field V_H on X (6.3)

A smooth real-valued function H on X
such that:

$$dH = V_H \rfloor \omega \qquad (6.4)$$

Definition. A **mechanical system** is a quadruple:

$$\{X, \omega, V_H, H\} \qquad (6.5)$$

satisfying 6.1-6.4. The **trajectories** of the mechanical system 6.5 are the orbit curves in X of the vector field V_H.

Theorem 6.1. Suppose given such a mechanical system. Let (V_f, f) be an element of the associated Poisson-Cartan Algebra $\mathbf{P}(\omega)$ such that:

$$\{(V_f, f), (V_H, H)\} = 0. \qquad (6.6)$$

Then, the one-parameter pseudogroup $\{s \rightarrow \exp(sV_f)\}$ generated by the vector field V_f is a symmetry group of the trajectories of the mechanical system 6.5 in the sense that the following condition is satisfied:

If $\{t \rightarrow \exp(tV_H)x_0\}$ is a trajectory, then, (6.7)
for each s, the curve $\{t \rightarrow \exp(sV_f)\exp(tV_H)x_0\}$
is also a trajectory.

Proof. We must prove that, for each t,

The tangent vector to the curve (6.8)
$\{t \rightarrow \exp(sV_f)\exp(tV_H)x_0\}$ is equal
to $V_H(\exp(sV_f)\exp(tV_H)x_0)$.

This follows from standard facts about the Jacobi-Lie Bracket [,] on vector fields and the relation:

$$[V_f, V_H] = 0 \tag{6.9}$$

which follows from 6.6.

q.e.d.

Of course, there is much, much more to be said about the relations between 'symmetry' of the equations of motion of mechancial systems and the algebraic stuctire of the Poisson-Cartan Algebra. This is related to much in the contemporary Applied iterature associated with the term **'Moment Map'**.Some such material has been covered in previous Volumes, particularly Vols. 17 and 23. Since this topic is not a main theme of this Volume, I will leave it at this point. Howvever, I would like to point out that much remains to be done to extend 'symmetry' and 'moment map' ideas to this Poisson-Cartan situation!

7. Critical points of functions whose Poisson Bracket with the Hamiltonian vanishes and orbits of one-parameter symmetry transformation groups which are also extremals of the mechanical system.

As a general bit of 'geometric philosophy', we know that studying the 'singularities' of a Geometric Structure is often of considerable Pure and Applied importance. For example, the contemporary theory of Dynamical Systems is dominated by the importance of the notion of an 'attractor', which is often such 'singularity' of the vector field generating the Dynamical System. The structure of Singularities is often closely related to underlying 'topological' data. A typical mathematical example of this relation is the **Morse Theory** of indices of singularity of real-valued smooth functions on a manifold.

Let us then consider an element (V, f) of the Poisson-Cartan algebra associated with a closed 2-form ω on a manifold X from this 'singularity' point of view. A **critical point** of the element f of $\mathbf{F}(X)$ is a point x of X such that:

$$df(x) = 0 \tag{7.1}$$

V and f are, by definition, linked by the following relation:

$$df = V_f \lrcorner \omega \tag{7.2}$$

We see that:

$$df(x) = 0' \text{ iff. } V_f(x) \lrcorner \omega = 0 \tag{7.3}$$

We have then proved:

Theorem 7.2. With the above notations, the point x of X is a critical point of the function 'f' if and only if the following condition is satisfied:

$$V_f(x) \text{ is a Cauchy characteristic vector of the 2-form } \omega. \tag{7.4}$$

As a typical 'application' of this point of view, I mention only the following:

Theorem 7.3. Suppose that the hypotheses of Theorem 7.2. Suppose in addition that the following 'non-singularity-of-the-Cauchy-Characteristic-vectors' is satisfied:

$$\text{The Cauchy Characteristic vectors } \{\mathbf{C}(\omega, x)\} \text{ are of constant dimension as x ranges over X.} \tag{7.5}$$

Then, for a point 'x_0' of X which is a critical point of 'f' the following condition is satisfied:

$$\text{The orbit curve } \{t \longrightarrow \exp(tV_f)(x_0)\} \text{ is a Cauchy Characteristic curve of } \omega. \tag{7.6}$$

Proof. We know that the vector field V_f is an infinitesimal symmetry of the Cauchy Characteristic vector field system, i.e.

$$V_f(\mathbf{C}(\omega)) \subset \mathbf{C}(\omega) \tag{7.7}$$

Hence, the one-parameter pseudogroup $\{t \to \exp(tV_f)\}$ generated by the vector field V_f maps the leaves of the Cauchy Characteristic foliation into themselves. By Theorem 7.2 and the hypothesis that 'x_0' is a critical point of f' implies that:

$$V_f(x_0) \, \varepsilon \, \mathbf{C}(\omega, x_0) \qquad (7.8)$$

i.e.

$V_f(x_0)$ is a tangent at x_0 to the Cauchy (7.9)
Characteristic foliation passing through x_0

Then, as proved in [6], we have:

V_f is tangent to the Cauchy (7.10)
Characteristic foliation passing through x_0
at each of its points.

(This is an easy consequence of the uniqueness theorem for solutions of ODE's with given initial conditions.) 7.10 proves 7.6.

q.e.d.

Remark. The reader will find much more about the properties of these 'trajectories of dynamical systems which are also **'orbits of one-parameter groups of symmetries'** in my papers [7, 8, 9]. Again, I must leave the further study of this situation to another point.

Bibliography for Mechanics

1. R. Abraham and J. E. Marsden, **Foundations of Mechanics**, 2nd Edn., Addison-Wesley, Reading, MA 1979

2. E. Cartan, **Lecons sur Les Invariants Integraux**, Hermann, Paris, 1922.

3. C. Caratheodory, **Calculus of Variations and Partial Differential Equations of the First Order,** vol. II, Holden-Day, 1967

4. C. Caratheodory, . a).Die Methode der geodatische Aquidistanten und das Probleme von Lagrange, *Acta Math.*, 47 (1926), 199-236., or *Gesammelten Mathematische Schriften*, Bd. I, 212-248. b). Uber die Einteilung der Variationsprobleme nach Klassen, Comment. Math. Helv. 5 (1933), 1-19., or Gesammelten Mathematische Schriften, Bd. I, 270-288.

5. R. Hermann, On Geodesics that are also Orbits, **Bull. Amer. Math. Soc.** 66 (1960), pp. 91-94.

6. R. Hermann, The differential geometry of foliations, part 2, *J. Math. Mech.* (Indiana Math J.), **11**, 303- 316 (1962).

7. R. Hermann, C-W Cell Decomposition of Symmetric Homogeneous Spaces, **Bull. Amer. Math. Soc.** 66 (1960), pp. 126-128.

8. R. Hermann, Some Differential Geometric Aspects of the Lagrange Variational Problem, **Illinois J. Math.** 6 (1962), pp. 634-674.

9. R. Hermann, E. Cartan's Geometric Theory of Partial Differential Equations, **Advances in Math** 1 (1965), pp. 265-316.

10. R. Hermann, Remarks on the Geometric Nature of Quantum Phase Space, **J. of Math. Phys.** 6 (1965), pp. 1768-1771.

12. R. Hermann, The Second Variation for Variational Problems in Canonical Form, **Bull. Amer. Math. Soc.** Vol 71 (1965), pp. 145-148.

13. R. Hermann, The Second Variation for Minimal Submanifolds, **J. of Math. and Mech.** 16 (1966), pp. 473-492.

14. R. Hermann, **Differential Geometry and the Calculus of Variations,** Academic Press New York 1969. Second Edition, Math Sci Press, 1976.

15. R. Hermann, Quantum Field Theories with Degenerate Lagrangians, **Phys. Rev.** 177 (1969) P. 2454.

16. R. Hermann, **Lie Algebras and Quantum Mechanics**, W. A. Benjamin New York 1971. 400 pp.

17. R. Hermann, **Vector Bundles in Mathematical Physics**, Parts I and II, W. A. Benjamin, New York 1970. 441 pp. and 400 pp.

18. R. Hermann, **Lectures on Mathematical Physics**. Vol. 1 W. A. Benjamin New York. 1970. 470 pp.

19. R. Hermann, Spectrum-generating Algebras in Classical Mechanics I and II. **J. Math. Phys.** 14. (1972) 873 - 878.

20. R. Hermann, Left Invariant Geodesics and Classical Mechanics on Manifolds, **J. Math. Phys.** 13 (1972) p. 460.

21. R. Hermann, **Geometry Physics and Systems,** Marcel Dekker New York 1974.

22. R. Hermann, **Geometric Structure Theory of Systems-Control Theory and Physics,** Part A, Vol. IX of Interdisciplinary Mathematics Math Sci Press, Brookline, Mass. 1976.

23. R. Hermann, **Gauge Fields and Cartan-Ehresmann Connections** , Part A Vol. X of Interdisciplinary Mathematics Math Sci Press, Brookline, Mass. 1976.

24. R. Hermann, Geodesics of Singular Riemannian Metrics. **Bull. Amer. Math. Soc.** 79 (1973), pp. 780-782.

25. R. Hermann, **Toda Lattices, Cosymplectic Manifolds, Baecklund Transformations and Kinks,** Part A and B, Math Sci Press, Brookline, MA 1976. Interdisciplinary Mathematics, vols. 15 and 18.

26. R. Hermann, **Quantum and Fermion Differential Geometry**, Part A, Math Sci Press, Brookline, Mass., 1976. Interdisciplinary Mathematics, vol. 16.

27. R. Hermann, **Yang-Mills, Kaluza Klein and the Einstein Program**, Interdisciplinary Mathematics, vol. 19, Math Sci Press, Brookline, Mass.

28. R. Hermann, Appendix: "Kleinian mathematics from an advanced standpoint", **Development of Mathematics in the 19th Century**, by Felix Klein, Math Sci Press, 1979.

29. R. Hermann, **Cartanian Geometry, Nonlinear Waves, and Control Theory**, parts A and B, Interdisciplinary Mathematics vols. 20 and 21., Math Sci Press, 1979.

30. R. Hermann, **Hamilton-Jacobi-Lie Theory and the Calculus of Variations**, Report, Math. Dept., Univ. of California, Berkeley, 1962.

31. R. Hermann, The Differential geometric Structure of General Mechanical Systems from the Lagrangian Point of View, **J. Math. Phys.** 23 (1982), 2077-2089.

32. R. Hermann, Geometric Theory of Deformation and Linearization of Pfaffian Systems and Its Application to System Theory and Mathematical Physics, **J. Math Phys.** 24, 2268-2276, 1984.

33. R. Hermann, The Geometric Foundations of the Integrability Property of Differential Equations, I: Lie's "Function Groups"., **J. Math Physics**, 24, 1983, 2422-2432; part II:Mechanics on Affinely Connected Manifolds and the work of Kowalewska amd Painleve, **J. Math Physics**, 25m 1984, 778-786.

34. R. Hermann,. **Topics in The Geometric Theory of Integrable Systems**, Math Sci Press, 1984.

35. R. Hermann,. **Topics in Physical Geometry,** Math Sci Press, 1988

36. R. Hermann, Differential Form Methods in the Theory of Variational Systems and Lagrangian Field Theories , **Acta App. Math.,** 12, 1988, 35-78.

37. R. Hermann, Lax Representation as a "Quantization" of the "Function Groups" of Sophus Lie, **Phys. Rev. D,** 26 , 1982, 1491-1492.

38. R. Hermann, Geometric Construction and Properties of Some families of Solutions of Nonlinear Partial Differential Equations, **J. Math. Phys.,** 24, 1983, 510-521.

39. R. Hermann, Geometric and Lie-Theoretic Principles in Pure and Applied Deformation Theory, in: M. Gerstenhaber and M. Hazewinkel (eds.), **Deformations of Algebras and Applications,** D. Reidel, Dordrecht, 1988, 701-796.

40. R. Hermann, **Geometric Structures and Nonlinear Physics,** Math Sci Press, 1992

41. R. Hermann, **Constrained Mechanics and Lie Theory,** Math Sci Press, 1993,

42. R. Hermann,The Differential Geometry of Foliations, I, **Annals of Math.,** 1961.

43. R. Hermann, Totally Geodesic Orbits of Groups of Transformations, **Proc. Ned. Akad. Wet.** 65 (1962), pp. 291-298.

44. R. Hermann, Perturbation theory for nonlinear feedback control systems and Spencer-Goldschmidt Integrability of linear partial differential equations., **Acta App. Math.,** 18, 17-57, 1990

45. R. Hermann, Differential Form Methods in the Theory of Variational Systems and Lagrangian Field Theories , **Acta App. Math.,** 12, 1988, 35-78.

46. R. Hermann and A. Krener, Nonlinear controllability and observability, **IEEE Trans. Aut. Contr.**, AC-22, 728-740 (1977).

47. R. Montgomery, Isoholonomic problems and some applications, **Commun. Math. Phys.**, 128, 565-592, 1990.

48. M. Morse, **Calculus of Variations in the Large**, Amer. Math. Soc.

49. D. J. Saunders, **The Geometry of Jet Bundles,** Cambridge Univ. Press, 1989.

50. E. Whittaker, **Analytical Dynamics of Particles and Rigid Bodies**, Cambridge University Press, 1966.

CHAPTER 5   REGULARIZATION OF SINGULARITIES OF ORDINARY DIFFERENTIAL SYSTEMS

My long-range goal is to develop 'geometric' ways of studying the (possibly) 'chaotic' long-time behavior of some of the equations of Celestial Mechanics, as uncovered in the Numerical work of Gerry Sussman and Jack Wisdom. As a preliminary to this, it seems to me to be worthwhile to understand better the one case where the 'long-time behavior' of orbits of 1 and 2-Bodies can be understood more precisely, namely that of collisions. This Chapter begins this study.

1. Prolongations of ODE systems.

The work of Levi-Civita, Kustaanheimo and Stiefel [3] on Regularization of the singularities of the differential equations of Celestial Mechanics suggests a general geometric setting in terms of Exterior Differentiable Systems. I will now pursue this goal.

Let us first use the development in the treatise by E. L. Stiefel and G. Scheifele [3] to formulate a more general geometric setting than the one they utilize. Suppose given the following data:

    Smooth manifolds X and Y   (1.1)

    A smooth mapping $\pi: Y \to X$ (1.2)

    Systems of ODE's for curves $\{t \to x(t)\}$ and $\{s \to y(s)\}$ on X and Y, respectively:

$$f(x, dx/dt, \ldots) = \text{constant} \qquad (1.3)$$

$$g(y, dy/ds, \ldots) = \text{constant} \qquad (1.4)$$

Remark. The ODE systems 1.3 may have 'singularities' on X. Precisely, the function 'f' on the left-hand side of 1.3 is a smooth mapping from an open subset of the space $J^r(R, X)$ of r-th order jets (with: r = order of the ODE system 1.3) to another manifold 'Z' and 'g' is a smooth mapping from an open subset of the space $J^r(R, Y)$ of s- order jets (with: s = order of the ODE system 1.4) to another manifold 'W'.

Definition. The mapping π is a **prolongation** from the system 1.3 to the system 1.4 if the following condition is satisfied:

> For every solution curve $\{s \to y(s)\}$ of the
> ODE system 1.4 the projected curve $\{s \to \pi y(s)\}$ \hfill (1.5)
> is, **up to a change in parameterization**, a solution
> of the ODE system 1.3.

Definition. The mapping π is a **regularization map** from the system 1.4 to the system 1.3 if it is a prolongation, in the sense that 1.5 is satisfied, and if the following condition is satisfied:

$$\text{'g' is a smooth mapping from } \textbf{all} \text{ of } J^s(R, Y) \text{ to W.} \quad (1.6)$$

It is very convenient for the purposes of studying prolongation and regularizations of ODE systems to define them in terms of Exterior Differential Systems. I will now briefly describe one setting for this.

2. ODE systems and their prolongations defined by systems of first-degree differential forms, i.e. Pfaffian Systems.

Let:

$$\{\theta^1, ..., \theta^m\} \quad (2.1)$$

be a collection of smooth one-forms on the manifold X.

ODE Regularization

Definition. A curve:

$$\varkappa: \{t \longrightarrow \varkappa(t)\} \qquad (2.2)$$

in X is said to be an **orbit curve** of the system associated with the collection 2.1 of 1-forms if the following condition is satisfied:

$$\theta^1(d\varkappa/dt) = 0 = \ldots = \theta^m(d\varkappa/dt) \qquad (2.3)$$

Remark. Here is another language for condition 2.3. Let:

$$\mathbf{ED}\{\theta^1, \ldots, \theta^m\} = \text{the Exterior Differential System} \qquad (2.4)$$
$$\text{generated by the 1-forms } \{\theta^1, \ldots, \theta^m\}.$$

Then, 2.3 is satisfied iff. the map $\varkappa$ of an interval of real numbers to X defining the curve $\{t \longrightarrow \varkappa(t)\}$ is an **integral manifold** of $\mathbf{ED}\{\theta^1, \ldots, \theta^m\}$ in the sense that:

$$\varkappa^d(\mathbf{ED}\{\theta^1, \ldots, \theta^m\}) = 0. \qquad (2.5)$$

Definition. Exterior Differential Systems which are generated in this way by one-forms are called **Pfaffian Systems.**

Let us now suppose that:

$$\dim X = n+1, \qquad (2.6)$$

$$(t, x^1, \ldots, x^n) \text{ are a coordinate system of} \qquad (2.7)$$
$$\text{smooth functions on X}$$

We can then expand the 1-forms $\{\theta_1, \ldots, \theta_m\}$ in terms of these coordinates 2.7 to obtain relations of the following form:

$$\theta^1 = \sum_{1 \le i \le n} f^1{}_i(x, t)dx^i + f^1(x, t)dt$$

$$\cdots \qquad (2.8)$$

$$\theta^m = \sum_{1 \le i \le n} f^m{}_i(x, t)dx^i + f^m(x, t)dt$$

Theorem 2.1. Let:

$$\{s \longrightarrow \aleph(s)\} \qquad (2.9)$$

be a curve in X. Let:

$$x(s) = (x^1(s) = x^1(\aleph(s)), ..., x^n(s) = x^n(\aleph(s)), t(s)) \qquad (2.10)$$

be its components in the coordinate system 2.7, considered as a curve in R^{n+1}. Then, the curve 2.9 is an orbit curve of the Pfaffian system $\{\theta^1, ..., \theta^m\}$ if and only if the coordinate functions 2.10 are a solution of the following ODE system:

$$\sum_{1 \le i \le n} f^1{}_i(x(s), t(s))dx^i/ds + f^1(x(s), t(s))dt/ds = 0 \qquad (2.11)$$

$$\cdots$$

$$\sum_{1 \le i \le n} f^1{}_i(x(s), t(s))dx^i/ds + f^1(x(s), t(s))dt/ds = 0 \qquad (2.12)$$

Proof. Follows from 2.3 and 2.8.

Let us now formulate the 'prolongation' concept in terms of Pfaffian systems.

Definition. Let $\{\theta^1, ..., \theta^m\}$ denote a Pfaffian system on the manifold X and let $\{\alpha^1, ..., \alpha^p\}$ denote a Pfaffian system on the manifold Y. Let $\pi: Y \to X$ denote a smooth mapping from Y to X. Then, π is said to be a **prolongation mapping from the Pfaffian systems** $\{Y, \alpha^1, ..., \alpha^p\}$ yo **the Pfaffian system** $\{X, \theta^1, ..., \theta^m\}$ if there are relations of the following form:

$$\pi^d(\theta^1) = \sum_{1 \le i \le p} A^1{}_i \alpha^i$$

$$\cdots \qquad (2.13)$$

$$\pi^d(\theta^m) = \sum_{1 \le i \le p} A^m{}_i \alpha^i$$

Another way of putting this is to require that:

$$\pi^d(\mathbf{ED}\{\theta^1, ..., \theta^m\}) \subset \mathbf{ED}\{\alpha^1, ..., \alpha^p\} \qquad (2.14)$$

Theorem 2.1. If π satisfies 2.13 or 2.14, then the map π has the following prolongation property:

π is a prolongation map between the ODE system
consisting of the orbit curves of the Pfaffian system $\qquad$ (2.15)
$\{\alpha^1, ..., \alpha^p\}$ to the orbit curves of the Pfaffian system
$\{\theta^1, ..., \theta^m\}$

Proof. Follows from 2.3 and the definition of 'π^d' as applied to 1-forms.

Let us now turn to study the prolongations associated with Hamiltonian systems, as developed in [3].

3. A method for Regularization of 1-D Hamiltonian Systems.

After these generalities, let us turn to the situations considered in [3]. In this Section, suppose given the following data:

$$X = R^3 = \text{space of variables } \{x, p, t\}. \tag{3.1}$$

$$(x, p, t) \longrightarrow H(x, p, t) \text{ is a real-valued function} \tag{3.2}$$
(possibly with singularities) on X, the **Hamiltonian.**

Set:

$$\omega = pdx - Hdt, \tag{3.3}$$
a one-form (again, possibly with singularities) on X.

Then:

$$d\omega = dp \wedge dx - dH \wedge dt \tag{3.4}$$

Set:

$$dH = H_x dx + H_p dp + H_t dt. \tag{3.5}$$

i.e. 'subscripts denote partial derivatives'.

The following results describe the relations between Exterior and Hamiltonian Systems. They are the fundamantal relations in Cartan's approach to Mechanics [2].

Theorem 3.1. The following formulas hold:

$$d\omega = dp \wedge dx - (H_x dx + H_p dp) \wedge dt \qquad (3.6)$$

$$= (dp + H_x dt) \wedge (dx - H_p dt) \qquad (3.7)$$

Let:

$C(d\omega)$ = exterior differential system generated by the one-forms: (3.8)

$$\{\alpha = dp + H_x dt, \; \beta = dx - H_p dt\} \qquad (3.9)$$

Then, a curve of the form:

$$\varkappa = \{t \rightarrow (x(t), p(t), t)\} \qquad (3.10)$$

in X is a Cauchy Characteristic Curve of $d\omega$, i.e. satisfies:

$$[d\varkappa/dt] \rfloor d\omega = 0 \qquad (3.11)$$

if and only if:

The curve $\{\{t \rightarrow (x(t), p(t))\}$ in R^2 satisfies Hamilton's equations 3.12-3.13:

$$dx/dt = H_p(x(t), p(t)) \qquad (3.12)$$

$$dp/dt = -H_x(x(t), p(t)) \qquad (3.13)$$

In turn, 3.12-3.13 are equivalent to the following condition:

The pull-back via the map $\varkappa$ of the one-forms α and β (defined in 3.9) is zero. (3.14)

Here is an another way of putting 3.14:

The curve ϰ defined by 3.10 is an orbit curve
of the Pfaffian system {α, β} defined by 3.9.

Proof. 3.6 follows from 3.5. 3.7 now follows after expanding the right handside of 3.7 and using the relation: $dt \wedge dt = 0$. The rest of the statements in the Theorem follows from the definitions and the algebraic rulkes satisfied by the operations '⌋' and '∧'.

q.e.d.

In [3], Stiefel and Scheifele provide ways in special cases (which go back to Euler) for regularizing 1-D Hamilton's Equations 3.12-3.13. Their method involves two steps:

Step 1: Change the time - parameterization of solutions of 3.12-3.13. (3.15)

Step 2: Change the x-dependence of the solutions of 3.12-3.13 so as to get rid of the 'singuarity'. (3.16)

Remark. Of course, the change of variable in Step 2 must also be 'singular, and it must be chosen so as to 'cancel out' the given singularity in the Hamiltonian function H.

We can formulate this in the following more uniform way:

Choose a one-form:

$$\theta = w^{-1}dt \qquad (3.17)$$

and a pair of functions on X

$$\{u, v\} \qquad (3.18)$$

such that the Cauchy Characteristic System of $d\omega$ is non-singular when written in terms of the basis $\{dy, dz, \theta\}$ for one-forms on X. 'w' is a new function on X to be chosen as a 'change of time parameterization'.

I will now describe the Algorithm for doing this utilized in [3].

4. The 1-D Regularization Algorithms in terms of a Moving Frame for the state space X.

Keep the notations of Section 3. Let us suppose that the basis $\{dx, dp, dt\}$ for 1-forms on X can be written in the following form:

$$dt = w\theta \qquad (4.1)$$

$$dx = x_u du \qquad (4.2)$$

$$dp = p_u du + p_v dv \qquad (4.3)$$

In 4.1-4.3, the right hand side is defined by the following conditions:

The $\{u, v, w\}$ are functions on X. $\qquad (4.4)$

θ is a 1-form on X $\qquad (4.5)$

$\{du, dv, \theta\}$ is a basis of one-forms on X, $\qquad (4.6)$
i.e. forms a **moving frame for X** in the sense
used extensively by Cartan.

Remarks. By a 'moving frame' for a manifold X of dimension n, we will mean a set:

$$\{\theta_1, ..., \theta_n\} \quad (4.7)$$

of one-forms on X such that the following condition is satisfied:

For each point x of X, the 1-covectors $\{\theta_1(x), ..., \theta_n(x)\}$ form a basis of the cotangent space X_x^d. (4.8)

The method being used here, (i.e. postulating a 'new moving frame' with certain properties which must be satisfied) is a standard one in Cartan's work. Of course, the 'moving frames' used in this 1-D situation are especially simple, and are not really necessary. Certainly, in their work Steifel and Scheifele [3] do not explicitly use Cartan's 'moving frame' ideas, although an attentive reader can perhaps detect a trace of them.

Insert relations 4.1-4.3 into 3.3:

$$\omega = px_u du - Hw\theta \quad (4.9)$$

Let us suppose also that 'v' is chosen so that:

$$px_u = v \quad (4.10)$$

Set:

$$H' = Hw \quad (4.11)$$

Combine 4.9 and 4.10:

$$\omega = vdu - H'\theta \quad (4.12)$$

We obtain another relation by exterior differentiating both sides of 4.1:

ODE Regularization

op:
$$0 = dw \wedge \theta + w d\theta,$$

$$d\theta = -w^{-1} dw \wedge \theta \qquad (4.13)$$

Remark. Relation 4.13 says that the Paffian System generated by $\{\theta\}$ is Frobenius Integrable.

Let us now exterior differentiate both sides of relation 4.12:

$$d\omega = dv \wedge du - dH' \wedge \theta + H' d\theta \qquad (4.14)$$

Define functions 'H'_u' and 'H'_v' by the following relation:

$$dH' \wedge \theta = [H'_u du + H'_v dv] \wedge \theta \qquad (4.15)$$

Combine 4.14 and 4.15:

$$d\omega = dv \wedge du - [H'_u du + H'_v dv] \wedge \theta - H' w^{-1} dw \wedge \theta \qquad (4.16)$$

Let us expand the one-form 'dw' in terms of the moving frame 4.6:

$$dw \wedge \theta = (w_u du + w_v dv) \wedge \theta \qquad (4.17)$$

Combine 4.16 and 4/17:

$$d\omega = dv \wedge du - [H'_u du + H'_v dv - H' w^{-1}(w_u du + w_v dv)] \wedge \theta \qquad (4.18)$$

Theorem 4.1. In terms of the new Moving Frame we have the following formula for the Symplectic Structure associated with the 2-form $d\omega$:

$$d\omega = [dv + (H'_u - H' w^{-1} w_u)\theta] \wedge [du - (H'_v + H' w^{-1} w_v)\theta] \qquad (4.19)$$

Proof. Multiply out the right-hand side of 4.15 and notice that it agrees with the right hand side of 4.14, using the relation '$\theta \wedge \theta = 0$'.

Theorem 4.2. Let:

$$\varkappa: \{s \longrightarrow \varkappa(s)\} \tag{4.20}$$

be a curve in X such that:

$$\varkappa^d(\theta) = ds \tag{4.21}$$

Set:

$$u(s) = \varkappa^d(u)(s) \tag{4.22}$$

$$v(s) = \varkappa^d(v)(s) \tag{4.23}$$

Then, $\varkappa$ is a Cauchy Characteristic curve of $d\omega$ if and only if the functions $\{u(s), v(s)\}$ defined by 4.22-4.23 satisfy the following differential equations:

$$du/ds = (H'_v + H'w^{-1}w_v) \tag{4.24}$$

$$dv/ds = -(H'_u - H'w^{-1}w_u) \tag{4.25}$$

Proof. Use formula 4.19.

Remark. Notice that the introduction of the 'non-holonomic' Moving Frame 'θ' in place of 'dt' has introduced the second terms on the right hand sides of 4.24 and 4.25 and destroyed the 'Hamiltonian' nature of the ODE system 4.24-4.25.

Let us now specialize the above formulas to the case of a particle of unit mass moving on the 1-D manifold 'R' in a central potential.

5. Regularization of a 1-D Newtonian particle.

ODE Regularization

Let us now specialize the material in Section 4 to the case treated in [3] of a unit mass particle moving in 1-D under an inverse-square central force. Thus:

$$\omega = pdx - (1/2p^2 + x^{-1})dt \qquad (5.1)$$

Set:
$$dt = x\theta \qquad (5.2)$$

$$x = u^2, \qquad (5.3)$$

leading to:
$$\omega = 2pudu - (1/2p^2 + u^{-2})u^2\theta \qquad (5.4)$$

Now, set:
$$v = 2pu, \qquad (5.5)$$

leading to:
$$\omega = vdu - (1/8v^2u^{-2} + u^{-2})u^2\theta$$

$$= vdu - (1/8v^2 + 1)\theta \qquad (5.6)$$

Exterior differentiate both sides of 5.2:

$$0 = dx \wedge \theta + xd\theta,$$

or:
$$d\theta = -x^{-1}dx \wedge \theta \qquad (5.7)$$

Exterior differentiate both sides of 5.6 and use 5.7 and 5,3:

$$d\omega = dv \wedge du - 1/4vdv \wedge \theta + (1/8v^2 + 1)u^{-1}du \wedge \theta \qquad (5.8)$$

Factor the right hand side of 5.8:

$$d\omega = (dv - (1/8v^2 + 1)u^{-1}\theta) \wedge (du - 1/4v\,\theta) \qquad (5.9)$$

Now, set

E = exterior differential system generated

by the 1-forms
$$\{dv - (1/8v^2 +1)u^{-1}\theta,\ du - 1/4v\,\theta\} \tag{5.10}$$

An orbit curve of **E** is then determined by a curve:

$$\{s \longrightarrow (u(s), v(s))\} \tag{5.11}$$

such that:

$$du/ds = 1/4v \tag{5.12}$$

$$dv/ds = (1/8v^2 +1)u^{-1}, \tag{5.13}$$

We also have the relation of 'conservation of energy', or:

$$1/8v^2 u^{-2} + u^{-2} = E \tag{5.14}$$

We can then substitute 5.14 into 5.13 to obtain the following system:

$$du/ds = 1/4v \tag{5.15}$$

$$dv/ds = uE, \tag{5.16}$$

5.15 and 5.16 together imply that:

$$d^2u/ds^2 = uE \tag{5.17}$$

5.17 is the differential equation of the Harmonic Oscillator, in accordance with the result of [3]. Let us summarize this work in the following way:

Theorem 5.1. Let:

$$\{t \longrightarrow (x(t), p(t))\} \tag{5.18}$$

be a solution of Hamilton's Equations, with the Hamiltonian H given by:

$$H = 1/2 p^2 + x^{-1} \tag{5.19}$$

Reparameterize the curve by 's' determined as follows:

$$s = \int x(t)^{-1} dt \tag{5.20}$$

Set:

$$u(s) = x(t(s))^2 \tag{5.21}$$

Then, the function $\{s \longrightarrow u(s)\}$ satisfies the linear, singularity-free Harmonic Oscillator Equation 5.17.

6. Reformulation of the 1-D Newtonian regularization in terms of symplectic structures and fiber bundles.

At the end of the book by Stiefel and Scheifele [3], they mention relations to differential geometry and fiber bundle theory. We can link the above results to these ideas by rephrasing them in the following form:

Theorem 6.1. Let:

$$Q = R - (0) \tag{6.1}$$

be the configuration space of a particle of unit mass moving on the real-line (minus the origin) according to the Inverse-Square Law. Set:

$$T^d(Q) = \text{cotangent bundle to } Q \tag{6.2}$$

$$H = \text{the real-valued function on } T^d(Q) \text{ defined by 5.19.}$$

Define a map:

$$\pi: T^d(Q) \longrightarrow T^d(Q) \tag{6.3}$$

by the following formula:

$$\pi((v, u)) = (v/2u, u^2). \tag{6.4}$$

Set:
$$H' = \pi^d(H). \tag{6.5}$$

For each real number 'E' set:

$$T^d(Q)(E) = \{(v, u): H'(v, u) = E\} \tag{6.6}$$

Let:
$$V_{H'} = \text{the infinitesimal symplectic vector field} \tag{6.7}$$
$$\text{on } T^d(Q) \text{ generated by } H'.$$

Then:

$$V_{H'} \text{ is tangent to each fixed energy submanifold } T^d(Q)(E). \tag{6.8}$$

$T^d(Q)(E)$ is an open subset of the following manifold:

$$\{(v, u): v, u \in R; H'(v, u) = E\} \qquad (6.9)$$

For each E, $V_{H'}$ extends smoothly to the manifold (6.10)
defined by 6.9.

The orbit curves of $V_{H'}$ on the submanifold 6.9 (6.11)
project down via the map 6.4 to the reparameterized
trajectories of the mechanical system moving on x-space
under the inverse-square law.

7. Final remarks.

As Stiefel and Scheifele point out in [3], these simple observations about regularization of the 1-D case point the way towards the study of more complicated systems. The map 'π' defined by 6.4 is a fiber bundle map, with Structure Group Z_2, the abelian group with two elements. The link to the 3-D case is that π must then be a fiber bundle mapping between manifolds of different dimension. Stiefel and Scheifele study in detail the 3-D, 2-Body Newtonian case, and study its implications for Numerical Analysis. They also point out that the fiber bundle map π which does the job involves the quaternions and is also the map which is utilized for the definition of the 'Hopf Map' between the 3-sphere and the 2-sphere.

I believe that there remains much to be done to study the differential-geometric foundations of this Regularization problem, particularly utilizing Cartan's Method of the Moving Frame. I hope to get back to this at a later point.

Bibliography

1. R. Bryant, S. Chern, R. Gardner, H. Goldschmidt and p. Griffiths, **Exterior Differential Systems**, Springer-Verlag, 1991

2. E. Cartan, **Lecons sur les Invariants Integraux**, Hermann, Paris, 1946.

3. E. L. Stiefel and G. Scheifele, **Linear and Regular Celestial Mechanics**, Springer-Verlag, 1971

4. K. Yang, **Exterior Differential Systems and Equivalence Problems**, Kluwer, 1992.

CHAPTER 6 - SOME LIE-THEORETIC ASPECTS OF THE 2-BODY PROBLEM OF CELESTIAL MECHANICS

1. Introduction.

The Numerical Analysis of n-Body Problems of Celestial Mechanics is a subject which is undergoing extensive development and modernization, utilizing today's extended computational power and techniques of differential geometry and Lie theory (symplectic structures, symplectic integrators, Lie series, ...). I believe that the special group-theoretic structure of the Kepler 2-Body Problem has a role to play in these developments. Ever since I stumbled on it while writing my book "Lie Groups for Physicists" in the 1960's, I have had a special fondness for the Lie-theoretic treatment of the Quantum- Mechanical, Non-Relativistic Hydrogen Atom developed by Pauli and Bargman in the 1930's. At the Classical-Mechanics Level, this involves an unexpectedly large symmetry algebra of the Newtonian 2-Body Hamiltonian, involving both the 3-dimensional symmetry of **Angular Momentum** and the 'extra' 3-dimensional symmetry given by the so-called **Runge-Lenz Vector.** Together, they generate an SO(4, R)-symmetry of the Hamiltonian on the positive-energy part of the 2-Body phase space.

This type of 'extra' symmetry should also be useful for Numerical purposes. One clue in this direction is the Kustaanheimo - Stiefel Regularization of the 3-D Newtonian 2-Body Problem, which involves the Quaternions. Now, the Quaternions form a 4-dimensional real vector space on which the orthogonal matrix group SO(4, R) acts. In this Chapter, I will review certain features of the Runge-Lenz Symmetry which seem potentially useful for Numerical purposes.

2. The 2-Body Problem of Newtonian Celestial Mechanics and the Runge-Lenz Vector.

Suppose given the following data:

$$X = \{(\mathbf{p}, \mathbf{x}): \mathbf{p} \in R^3; \mathbf{x} \in R^3 - 0\} \tag{2.1}$$
= the state space of a particle moving in R^3, with a singularity at the origin.

$$H = 1/2|\mathbf{p}|^2 - 1/r = \text{Hamiltonian of Kepler Problem} \tag{2.2}$$

Remark. $\{\mathbf{p} \rightarrow |\mathbf{p}|\}$ is the usual Euclidean norm for vectors in R^3.

$$r^2 = |\mathbf{x}|^2 \tag{2.3}$$

Choose indices as follows:

$$1 \leq i, j, k, ..., \leq 3. \tag{2.4}$$

$\{x^i\}$ and $\{p^i\}$ are the component of the vectors $\mathbf{x}$ and $\mathbf{p}$, respectively.

They are smooth functions on the symplectic manifold X.

The Standard Symplectic Form on X is:

$$d\mathbf{p} \wedge d\mathbf{x} = \Sigma_i dp^i \wedge dx^i \tag{2.5}$$

Set:

$$f^{ij} = x^i p^j - x^j p^i \tag{2.6}$$

$$f^i = x^i/r - x^i |\mathbf{p}|^2 + p^i(\mathbf{p} \cdot \mathbf{x}) \tag{2.7}$$

$\{f^{ij}\}$ are the **angular momenta** and the $\{f^i\}$ are what the physicists call the **Runge-Lenz Vector**. The following Poisson Bracket commutation relations hold:

$$\{f^{ij}, H\} = \{f^i, H\} = 0 \tag{2.8}$$

$$\{f^i, f^j\} = 2Hf^{ij} \tag{2.9}$$

$$\{f^{ki}, f^j\} = \delta^{ji}f^k - \delta^{jk}f^i \tag{2.10}$$

$$\{f^{ij}, f^{kl}\} = \delta^{jk}f^{il}p^j - \delta^{ik}f^{jl} - \delta^{jl}f^{ik} + \delta^{il}f^{jk} \tag{2.11}$$

2.8 says that:

> The (f^{ij}, f^k) are conserved under the symplectic flow on X generated by the Hamiltonian H.

In addition, the following condition holds in the algebra of smooth, real-valued functions on X:

$$H\Sigma_{ij}f^{ij}f^{ij} + \Sigma_i f^i f^i = 1 \tag{2.12}$$

or:

$$H = \left[1 - \Sigma_i f^i f^i\right]\left[\Sigma_{ij}f^{ij}f^{ij}\right]^{-1} \tag{2.13}$$

These relations say that the system composed of H and the $\{f^{ij}, f^i\}$ forms a **Function Group** in the sense of Lie. (See explanations in Vols. 27 and 28, plus "Geometry, Physics and Systems".) I will now describe this data in terms of finite dimensional Lie algebras.

3. The description of the Kepler Hamiltonian and its symmetries in terms of finite dimensional Lie algebras and Lie pseudogroups.

Keep the notations of section 2. I will now rewrite the commutation relations 2.8-2.11 in a form which involves constant coefficients, i.e. the action of a finite dimensional real Lie algebra. Set:

$$X^+ = \{(\mathbf{p}, \mathbf{x}) \in X: H(\mathbf{p}, \mathbf{x}) > 0\} \tag{3.1}$$

$$X^- = \{(\mathbf{p}, \mathbf{x}) \in X: H(\mathbf{p}, \mathbf{x}) < 0\} \tag{3.2}$$

$$X^0 = \{(\mathbf{p}, \mathbf{x}) \in X: H(\mathbf{p}, \mathbf{x}) = 0\} \tag{3.3}$$

In X^+, set:

$$F^i = f^i/(H)^{1/2} \tag{3.4}$$

In X^-, set:

$$F^i = f^i/(-H)^{1/2} \tag{3.5}$$

We now have:

Theorem 3.1. In the open subset X^+ of X, the following Poisson Bracket commutation relations hold:

$$\{f^{ij}, H\} = \{F^i, H\} = 0 \tag{3.6}$$

$$\{F^i, F^j\} = 2f^{ij} \tag{3.7}$$

$$\{f^{ki}, F^j\} = \delta^{ji}F^k - \delta^{jk}F^i \tag{3.8}$$

$$\{f^{ij}, f^{kl}\} = \delta^{jk}f^{il} - \delta^{ik}f^{jl} - \delta^{jl}f^{ik} + \delta^{il}f^{jk} \tag{3.8a}$$

In the open subset X^- of X, the following Poisson Bracket commutation relations hold:

$$\{f^{ij}, H\} = \{F^i, H\} = 0 \tag{3.9}$$

$$\{F^i, F^j\} = -2f^{ij} \tag{3.10}$$

$$\{f^{ki}, F^j\} = \delta^{ji}F^k - \delta^{jk}F^i \tag{3.11}$$

$$\{f^{ij}, f^{kl}\} = \delta^{jk}f^{il} - \delta^{ik}f^{jl} - \delta^{jl}f^{ik} + \delta^{il}f^{jk} \tag{3.11a}$$

In X^+ (resp. X^-) the Poisson Bracket commutation relations given above say that:

> The smooth, real-valued functions (f^{ij}, F^k) define a finite dimensional Lie subalgebra of the Lie algebra (under Poisson Bracket) of all smooth, real-valued functions on X^+ (resp. X^-).

Proof. Follows from 2.6-2.8 and the following general relation between Poisson Bracket: $(f, f') \to \{f, f'\}$ and Scalar Multiplication $(f, f') \to ff'$ on real-valued functions on X:

$$\{f, f'f''\} = \{f, f'\}f'' + f'\{f, f''\} \tag{3.12}$$

Let us now translate these Poisson Bracket relations into Jacobi-Lie Bracket relations among vector fields on X^+ and X^-

Set:
$$V^H = \Sigma_k [\partial H/\partial p^k] \partial/\partial x^k - [\partial H/\partial x^k] \partial/\partial p^k \tag{3.13}$$

$$V^i = \Sigma_k [\partial F^i/\partial p^k] \partial/\partial x^k - [\partial F^i/\partial x^k] \partial/\partial p^k \tag{3.14}$$

$$V^{ij} = \Sigma_k [\partial f^{ij}/\partial p^k] \partial/\partial x^k - [\partial f^{ij}/\partial x^k] \partial/\partial p^k \tag{3.15}$$

$$[V^{ij}, V^{kl}] = \delta^{jk}V^{il} - \delta^{ik}V^{jl} - \delta^{jl}V^{ik} + \delta^{il}V^{jk} \qquad (3.15a)$$

Theorem 3.2. On X^+ the vector fields defined by 3.13-3.15 satisfy the following Jacobi - Lie Bracket commutation relations.

$$[V^{ij}, V^H] = [V^i, V^H] = 0 \qquad (3.16)$$

$$[V^i, V^j] = 2V^{ij} \qquad (3.18)$$

$$[V^{ki}, V^j] = \delta^{ji}V^k - \delta^{jk}V^i \qquad (3.19)$$

$$[V^{ij}, V^{kl}] = \delta^{jk}V^{il} - \delta^{ik}V^{jl} - \delta^{jl}V^{ik} + \delta^{il}V^{jk} \qquad (3.19a)$$

Similiarly, On X^- the vector fields defined by 3.13-3.15 satisfy the following Jacobi - Lie Bracket commutation relations.

$$[V^{ij}, V^H] = [V^i, V^H] = 0 \qquad (3.20)$$

$$[V^i, V^j] = -2V^{ij} \qquad (3.21)$$

$$[V^{ki}, V^j] = \delta^{ji}V^k - \delta^{jk}V^i \qquad (3.22)$$

$$[V^{ij}, V^{kl}] = \delta^{jk}V^{il} - \delta^{ik}V^{jl} - \delta^{jl}V^{ik} + \delta^{il}V^{jk} \qquad (3.22a)$$

Proof. We know that the assignment: $f \longrightarrow V$ such that: $df = V_f \rfloor \omega$, where: ω = the symplectic 2-form 2.5, is a Lie algebra homomorphism from $F(X)$ (made into a Lie algebra by means of the Poisson Bracket operation $\{ , \}$) to the Lie algebra (under Jacobi-Lie Bracket) of all smooth vector fields $F(X)$. Using this fact, we see that the Jacobi- Lie bracket relations 3.16-3.22 follow from the corresponding Poisson Bracket relations 3.6-3.11a.

q.e.d.

Remark. The **2-Body Problem** is to find the orbits of the vector field V^H. The **Regularization Problem** is to describe geometrically what happens at the singularity point $\{x = 0\}$. The above commutation relations enable us to say much of a qualititative, Lie-theoretic nature about these questions.

4. The commutation relations on the energy surfaces. 'Deformation' of the Lie algebras defined by the Runge-Lenz vectors as a smooth function of the energy.

The way that we have defined the vector fields $\{V^{ki}, V^j\}$ on X^+ and X^-, so as to define different Lie algebras as one passes from positive to negative energy is misleading fron the point of view of Deformation Theory, since there seems to be a discontinuity as one jumps from X^+ to X^-. However, alternate definitions are possible, which imply a perfectly smooth transition as one passes from positive to negative energy, as required for the Nijenhuis-Richardson Theory of Deformation of Lie Algebras. (See Ref. 4 of Chapter 1). Start again with the unnormalized Poisson Bracket relations 2.8-2.10:

$$\{f^{ij}, H\} = \{f^i, H\} = 0 \tag{4.1}$$

$$\{f^i, f^j\} = 2Hf^{ij} \tag{4.2}$$

$$\{f^{ki}, f^j\} = \delta^{ji}f^k - \delta^{jk}f^i \tag{4.3}$$

$$[f^{ij}, f^{kl}] = \delta^{jk}f^{il} - \delta^{ik}f^{jl} - \delta^{jl}V^{ik} + \delta^{il}f^{jk} \tag{4.3a}$$

Set:

V^{ij} = vector field on X corresponding to f^{ij} defined by the symplectic form 2.5. (4.4)

W^i = vector field on X corresponding to f^i defined by the symplectic form 2.5. (4.5)

V_H = vector field on X corresponding to H defined by the symplectic form 2.5. (4.6)

Then, we have (using 28-2.10 and 3.12) the following Jacobi- Lie Bracket commutation relations:

$$[V^{ij}, V^H] = [W^i, V^H] = 0 \tag{4.7}$$

$$[W^i, W^j] = 2V^H f^{ij} + 2HV^{ij} \tag{4.8}$$

$$[V^{ki}, W^j] = \delta^{ji}W^k - \delta^{jk}W^i \tag{4.9}$$

$$[V^{ij}, V^{kl}] = \delta^{jk}V^{il} - \delta^{ik}V^{jl} - \delta^{jl}V^{ik} + \delta^{il}V^{jk} \tag{4.9a}$$

Now, the commutation relations 4.7-4.9a do not define a finite dimensional Lie Algebra of vector fields on X. However, we can then restrict these vector fields to the 'energy surfaces' in X to obtain a continuous family of finite dimensional Lie Algebras.

Definition. For each real number E, set:

$$X(E) = \{(\mathbf{p}, \mathbf{x}): H(\mathbf{p}, \mathbf{x}) = E\} \tag{4.10}$$

X(E) is called the **Energy Surface** for the Hamiltonian H.

Let us first prove a general result about infinitesimally symplectic vector fields and Poisson Brackets on submanifolds of symplectic manifolds.

Theorem 4.1. Let $\{X, \omega\}$ be a smooth symplectic structure on the manifold X. Let:

$$H: X \longrightarrow R \qquad (4.11)$$

be a smooth, real-valued function on X. Let E be a real number. Set:

$$X(H, E) = \{x \in X: H(x) = E\} \qquad (4.12)$$

Let us suppose that the following condition is satisfied:

$$dH(x) \neq 0 \text{ for all } x \in X(H, E). \qquad (4.13)$$

Then:
$$X(H, E) \text{ is a smooth submanifold of X of codimension 1} \qquad (4.14)$$

Suppose that f is an element of $\mathbf{F}(X)$ such that:

$$\{f, H\} = 0, \qquad (4.15)$$

where $\{\,,\,\}$ is the Poisson Bracket operation on $\mathbf{F}(X)$ associated with ω. Let:

$$V_f = \text{infinitesimally symplectic vector field associated} \qquad (4.16)$$
$$\text{with f via the relation: } df = V_f \rfloor \omega$$

Then:

$$V_f \text{ is tangent to the submanifold } X(H, E) \qquad (4.17)$$

If f, f' are two elements of $\mathbf{F}(X)$ such that
$$\{f, H\} = 0 = \{f', H\}, \qquad (4.18)$$
then:

$$\{\{f, f'\}, H\} = 0 \qquad (4.19)$$
and

$$V_{\{f, f'\}} \text{ restricted to } X(H, E) \quad (4.20)$$

is the Jacobi-Lie Bracket of the vector fields V_f and $V_{f'}$ restricted to $X(H, E)$.

Proof. 4.14 is a result of the standard Implict Function Theorem. 4.18 implies that: $V_f(H) = 0$, which implies 4.17. It is a general property of the Poisson Bracket that:

$$V_{\{f, f'\}} = [V_f, V_{f'}] \quad (4.21)$$

4.21 implies 4.20. q.e.d

Let us return to the study of the Kepler Problem.

Theorem 4.2. For each real number E, the vector fields $\{V^H, V^i, V^{ij}\}$ are tangent to each of the submanifolds $X(E)$ of X. Denote these vector fields on the manifold $X(E)$ as:

$$\{V_E^H, V_E^i, V_E^{ij}\}. \quad (4.21a)$$

The following commutation relations hold for the vector fields 4.21a) on the submanifold $X(H, E)$:

$$[V_E^{ij}, V_E^H] = [V_E^i, V_E^H] = 0 \quad (4.22)$$

$$[W_E^i, W_E^j] = 2EV_E^{ij} + 2f_E^{ij}V_E^H \quad (4.23)$$

$$[V_E^{ki}, W_E^j] = \delta^{ji}W_E^k - \delta^{jk}W_E^i \quad (4.24)$$

$$[V_E^{ij}, V_E^{kl}] = \delta^{jk}V_E^{il} - \delta^{ik}V_E^{jl} - \delta^{jl}V_E^{ik} + \delta^{il}V_E^{jk} \quad (4.25)$$

Proof. Follows from Theorem 4.1 and 4.7-4.9a.

Theorem 4.3. The vector fields:

$$\{V_E^{ij}, W_E^i, f_E^{ij}V_E^H, f_E^{ij}V_E^H\} \tag{4.26}$$

span a finite dimensional Lie subalgebra of $U(X(E))$.

Proof. Follows from 4.22-4.25.

Theorem 4.4. On the manifold $X(E)$ of constant energy E, set:

$$W_E^{ij} = 2EV_E^{ij} + 2f_E^{ij}V_E^H \tag{4.27}$$

Then, the vector fields

$$\{W_E^{ij}, W_E^i\} \tag{4.28}$$

span a finite dimensional Lie subalgebra of $U(X(E))$.

Proof. Left to the reader.

Remark. Notice how the Lie subalgebras of $U(X(E))$ spanned by 4.27 and 4.28 move 'smoothly' with E, just as required by the Nijenhuis-Richardson Theory. This 'deformation' is the Lie Algebra analogue of the Kodaira-Spencer framework for Deformation of Complex Structures. (In the Complex Structure case, the 'Complex Structure" is defined by a Lie Subalgebra of the Lie Algebra of all smooth complex-valued vector fields.)

Lie Groups and Kepler

5. The pseudogroup action of SO(4, R) on the positive-energy part of the phase space regularization of the Kepler trajectories.

Let us turn to the study of the finite dimensional Lie agebra of vector fields on the positive-energy part X^+ of the phase space 2.1 defined by the relations 3.13-3.22a).

Theorem 5.1. The Lie algebra L^+ of vector fields spanned by the vector fields $\{V^{ij}, V^i\}$ on X^+ is isomorphic to the Lie alagebra of the connected Lie group SO(4, R) of 4x4 real orthogonal matrices ofdeterminant one.

Proof. It is known from (finite dimensional) Lie Group Theory that:

$$\text{the Lie algebra of SO(4, R) is isomorphic} \tag{5.1}$$
$$\text{to the Lie algebra of 4x4 real, skew-symmetric matrices.}$$

Now, every 4x4 real, skew-symmetric matrix can be paritioned into a 3x3 skew-symmetric matrix and a 3x1 column matrix. It is readily see (exercise for the reader) that the commutation relations 3.16-3.19a of the Lie algebra L^+ result.

q.e.d.

Set:

$$G^+ = \text{the simply connected Lie group whose} \tag{5.2}$$
$$\text{Lie algebra is isomorphic to } L^+.$$

Now, G^+ does not act 'globally' on X^+ because the vector fields in L^+ are not complete. However:

$$G^+ \text{ does act 'locally, as a Lie pseudogroup, on } X^+. \tag{5.3}$$

Let us ask:

$$\text{How is the flow generated by } V^H, \text{ whose orbits define} \tag{5.4}$$

the time-evolution of Hamilton's Equations with Hamiltonian H given by 2.2, related to the action of the Lie algebra L^+ and the Lie pseudogroup G^+ on X^+?

Let us start to answer this question with relation 2.12:

$$H\Sigma_{ij} f^{ij} f^{ij} + \Sigma_i f^i f^i = 1 \tag{5.5}$$

Set:

$$J = \Sigma_{ij} f^{ij} f^{ij} \tag{5.6}$$

$$M = \Sigma_i f^i f^i \tag{5.7}$$

Then, 2.12 takes the following form:

$$HJ + M = 1. \tag{5.8}$$

If 'f' is a function on X, let 'V^f' denote the vector field on X such that:

$$V^f(f') = \{f, f'\} \text{ for all } f' \, \varepsilon \, F(X). \tag{5.9}$$

Then:

$$V^{(ff')} = fV^{f'} + f'V^f \tag{5.10}$$

Applying the identities 5.9 and 5.10 gives the following:

$$HJ + M = 1$$
$$JV^H + HV^J + V^M = 0,$$

or:

$$JV^H = (-V^M - HV^J) \tag{5.11}$$

We now have proved:

Theorem 5.2. Formula 5.11 expresses the Time-Evolution Equation of the Kepler 2-Body Hamiltonian 2.2 on X^+ in terms of the action of the Lie algebra of $SO(4, R)$ as a Lie subalgebra of $U(X^+)$ and as a Lie subalgebra of $F(X^+)$, with $F(X^+)$ made into a Lie algebra via the Poisson Bracket associated with the symplectic form 2.5.

We can now see in a qualitative way what this has to do with 'regularization'. V^H has a 'singularity' at the point '$x = 0$' of configuration space R^3. As we recall was the case in the 1-D Kepler Regularization (see also the treatment in [3-5]) the First Step in the Regularization is to 'change the parameterization of the orbit curves of V^H'. 5.11 suggests a way of doing this. Set:

$$W^H = JV^H \qquad (5.12)$$

W^H is a vector field on X, again, with 'singularities', but with 'better' behavior at '$x=0$' than V^H. Since W^H and V^H only differ by multiplication by a function, we have:

> The orbit curves of W^H are same as the orbit curves (5.13)
> of V^H up to a change-in-parameterization.

Using 5.11, we see that:

$$W^H = -(V^M + HV^J) \qquad (5.14)$$

5.4 expresses the following property:

Theorem 5.3. The orbit curves of W^H on X^+ are each the orbit curves of a vector field in the Lie algebra L^+.

Proof. Using 2.8, we have:

$$W^H(H) = JV^H(H) = 0 \tag{5.15}$$

$$W^H(f^i) = JV^H(f^i) = 0 \tag{5.16}$$

$$W^H(f^{ij}) = JV^H(f^{ij}) = 0 \tag{5.17}$$

$$VM = \Sigma_i f^i V^i \tag{5.18}$$

$$VJ = \Sigma_{ij} f^{ij} V^{ij} \tag{5.19}$$

we see from 5.14 that an orbit curve $\{t \longrightarrow x(t)\}$ of W^H on $X(E)$ is the parameter-reversed orbit curves of the following vector field in L^+:

$$\Sigma_i f^i(x(0))V^i + E\Sigma_{ij} f^{ij}(x(0))V^{ij} \tag{5.20}$$

q.e.d.

This result suggests the following strategy for Regularizing the orbit curves of W^H:

> Find a Prolongation of the Lie algebra L^+ of (5.21)
> vector fields on X, i.e. a manifold Z, a submersion map
> $\pi: Z \longrightarrow X$, a Lie algebra $L^+{}_Z$ of vector fields on Z such
> that:
>
> π_* maps $L^+{}_Z$ onto L^+, (5.22)
> i.e. each of the vector fields in $L^+{}_Z$
> is projectable under π and maps into a vector
> field in L^+.
>
> The vector fields in $L^+{}_Z$ are **complete**. (5.23)

The explicit construction of such a Z and map π is the main topic of ref. 5.

There is a 'general setting' for this form of Regularization in termsof what I have called **Lie Structures** in Volume 27 and 28. I plan to return to the study of this setting in a later work.

Bibliography

1. D. C. Heggie, The N-Body Problem in Stellar Dynamics, in **Long-Term Behavior of Natural and Artificial N-Body Systems**, Ed. A. E. Roy, Kluwer, 1988

2. R. Hermann, **Differential Geometry and the Calculus of Variations**, Academic Press, 1967; Second Edition, Math Sci Press, 1977.

3. T. Levi-Civita, Sur la regularization du probleme des trois corps, Acta Math., 42, (1918), 99-144

4. C. L. Siegel and J. K. Moser, Lecturs of Celestial Mechanics, Springer, 1971

5. E. L. Stiefel and G. Scheifele, **Linear and Regular Celestial Mechanics**, Springer, 1971

CHAPTER 7. CARTAN'S FORMULA FOR THE POISSON BRACKET GENERALIZED TO CONSTRAINED SYSTEMS.

1. Introduction.

When reading the work of Lie and Cartan today, it is amazing to realize how much they anticipated of the differential-geometric mathematics which is currently used in quantum and classical mechanics, pure and applied mathematics, particularly that tied up with the term 'Poisson Bracket'. Lie emphasized the theory of what he called 'Function Groups', and that are now seen to be closely linked to what we call 'Poisson Structures'. Cartan's choice of topics to work on must have been influenced by Lie's example, but he used his own methods, mainly involving **differential forms.** In this Chapter, I will adapt Cartan's ideas on Poisson Bracket and Symplectic Structures to the needs of 'Mechanics with Constraints' and discuss some typical situations where these methods are useful. The foundational doculments are Cartan's book "Lecons sur Les Invariants Integraux" and my 1962 paper "Some Differential Geometric Aspects of the Calculus of Variations" in the **Illinois Journal of Mathematics.**

2. Symplectic manifolds, Poisson Bracket and symmetries of vector field systems generated by infinitesimal symplectic automomorphisms.

Let us first recall standard Symplectic lore.

Definition. A **symplectic structure** for the manifold X is defined by a smooth 2-differential form ω on X satisfying the following conditions 2.1-2.2:

$$d\omega = 0, \text{ i.e., } \omega \text{ is a closed differential form.} \quad (2.1)$$

The Exterior Differential System generated by ω has no Cauchy Characteristic vectors, i.e. the following condition is satisfied: (2.2)

If $v \in T(X)$ is a tangent vector to X such that
$$v \rfloor \omega = 0,$$
then: (2.3)
$$v = 0.$$

Certain purely algebraic facts follow from 2.2-2.3:

Theorem 2.1. Suppose that ω is a 2-differential form on the manifold X which satisfies 2.2-2.3. Then:

$$X \text{ is even dimensional; say: } \dim(X) = 2n \quad (2.4)$$

$$\omega^n \neq 0 \quad (2.5)$$
(ω^n denotes the exterior product of n copies of ω)

Proof. For each $x \in X$, the value of ω at x, denoted as '$\omega(x)$', is a skew-symmetric, bilinear form on the tangent space to X at x denotes as: X_x. The standard linerar algebra 'canonical form' for such a bilinear form implies that there is a basis:

$$\{v_1, ..., v_m\} \quad (2.6)$$

for X_x such that:

$$\omega(x)(v_i, v_j) = 1 \text{ if } 1 \le i \le n \text{ and } j = n+i \quad (2.7)$$
$$= 0 \text{ otherwise}$$

We see that:

Condition 2.3 implies that:
$$m = 2n \tag{2.8}$$

and 2.9-2.11:
$$\{v_1, ..., v_{2n}\} \text{ is a basis for } X_x. \tag{2.9}$$

If $\{\theta^1, ..., \theta^{2n}\}$ is the basis of the dual space X_x^d which is dual to the basis $\{v_1, ..., v_{2n}\}$ for X_x, then: $\tag{2.10}$

$$\omega(x) = \sum_{1 \leq i \leq n} \theta^i \wedge \theta^{n+i} \tag{2.11}$$

Using the rules of exterior multiplication, we see that:

$$\omega(x)^n = \theta^1 \wedge ... \wedge \theta^{2n} \tag{2.12}$$

q.e.d.

3. The Poisson Bracket operation on smooth real-valued functions on a symplectic manifold.

Now, suppose that X has such a symplectic structure $\{\omega\}$. **F**(X) (the space of C^∞ real valued functions on X) has a Lie algebra structure, called **Poisson Bracket**, denoted by

$$\{f_1, f_2\}. \tag{3.1}$$

Recall its definition. If f_1 and f_2 are elements of **F**(X), let $V(f_1)$ be the vector field on X defined by the following formula:

$$V(f_1) \,\lrcorner\, \omega = df_1 \tag{3.2}$$

Then,

$$\{f_1, f_2\} = V(f_1)(f_2) \tag{3.3}$$

Theorem 3.1. The Poisson Bracket 3.3 satisfies the following relation:

$$\{f_1, f_2\} = \omega(V(f_2), V(f_1)) \tag{3.4}$$

Proof. Use 3.3:

$\{f_1, f_2\} = V(f_1)(f_2) = df_2(V(f_1)) = V(f_1) \lrcorner\, df_2 =$, using 3.2,

$V(f_1) \lrcorner\, V(f_2) \lrcorner\, \omega = [V(f_2) \lrcorner\, \omega](V(f_1)) = \omega(V(f_2), V(f_1))$

q.e.d.

Theorem 3.2. Formula 3.3 defines an R-bilinear operation

$$[(f_1, f_2) \longrightarrow \{f_1, f_2\}]:$$
$$\mathbf{F}(X) \times \mathbf{F}(X) \longrightarrow \mathbf{F}(X)$$

which is skew-symmetric and satisfies the Jacobi Identity, i.e. makes $\mathbf{F}(X)$ into a real Lie algebra. For f_1 held fixed, the map: $f_2 \longrightarrow \{f_1, f_2\}$ is a first order linear differential operator.

Proof. Skew-symmetry follows from 3.4. The 'first order linear differential operator' property follows from 3.2. The proof of the Jacobi Identity is left as an Exercise.

For comparison, let us derive the classical formula for Poisson Bracket

4. The classical formula for Poisson Bracket.

Let us now suppose that:

$$X = R^{2n} \tag{4.1}$$

$$= \{(p, q): p, q \in R^n, p = (p_1, ..., p_n) = \{p_i: 1 \leq i \leq n\}, \tag{4.2}$$
$$q = (q^1, ..., q^n) = \{q^i: 1 \leq i \leq n\}\}$$

$$\omega = \Sigma_i \, dp_i \wedge dq^i \tag{4.3}$$

For $f \in \mathbf{F}(X)$,

$$df = \Sigma_i \, [\partial f/\partial p_i]dp_i + \Sigma_i \, [\partial f/\partial q^i]dq^i \tag{4.4}$$

$$= V(f) \, \rfloor \, \omega = \Sigma_i \, V(f)(p_i)dq^i - \Sigma_i \, V(f)(q^i)dp_i \tag{4.5}$$

Comparing 4.4 and 4.5, we have proved:

Theorem 4.1. The following formula determines the symplectic vector field $V(f)$ assigned to the real-valued function f on $X = R^{2n}$:

$$\partial f/\partial p_i = V(f)(q^i) \tag{4.6}$$

$$\partial f/\partial q^i = -V(f)(p_i) \tag{4.7}$$

Let us combine 2.5 and 4.6-4.7 to compute the classical formula for Poisson Bracket. By the definition of the Lie derivative of a function by a vector field, we have:

$$V(f_1) = \Sigma_i \, V(f_1)(p_i)\partial/\partial p_i + \Sigma_i \, V(f_1)(q^i)\partial/\partial q^i \tag{4.8}$$

$$=, \text{using 4.6-4.7}, \Sigma_i \, -[\partial f_1/\partial q^i]\partial/\partial p_i + \Sigma_i \, [\partial f_1/\partial p_i]\partial/\partial q^i \tag{4.9}$$

Hence,

$$\{f_1, f_2\} = V(f_1)(f_2) = \sum_i -[\partial f_1/\partial q^i][\partial f_2/\partial p_i] + \sum_i [\partial f_1/\partial p_i][\partial f_2/\partial q^i]$$
(4.10)

Note that formula 4.10 is indeed the classical formula for the Poisson Bracket.

5. Cartan's formula for the Poisson Bracket in terms of exterior multiplication.

Cartan based his coordinate-free treatment of Poisson Bracket (in his book "Lecons sur Les Invariants Integraux") on a formula involving exterior multiplication '∧' rather than, as above, using the inner-product operation '⌋' between vector fields and differential forms. Let us first derive some algebraic relations between these operations:

Theorem 5.1. Let θ and θ' be two differential forms on X, of degree 'r' and 's' respectively. Let V be a vector field on X. Then:

$$V \,\rfloor\, (\theta \wedge \theta') = (V \,\rfloor\, \theta) \wedge \theta' + (-1)^r \theta \wedge (V \,\rfloor\, \theta')$$
(5.1)

Proof. It suffices to prove 5.1 in the case that θ and θ' are monomials in the exterior algebra, i.e. suppose that:

$$\theta = \omega_1 \wedge \omega_2 \wedge \ldots \wedge \omega_r$$

$$\theta' = \omega_1' \wedge \omega_2' \wedge \ldots \wedge \omega_s'$$

Then:

$$\theta \wedge \theta' = \omega_1 \wedge \omega_2 \wedge \ldots \wedge \omega_r \wedge \omega_1' \wedge \omega_2' \wedge \ldots \wedge \omega_s'$$

Hence:

$$V \,\rfloor\, (\theta \wedge \theta') = \omega_1(V) \omega_2 \wedge \ldots \wedge \omega_r \wedge \omega_1' \wedge \omega_2' \wedge \ldots \wedge \omega_r' - \omega_1 \omega_2(V) \wedge \ldots \wedge \omega_r \wedge \omega_1' \wedge \omega_2' \wedge \ldots \wedge \omega_s'$$
$$+ \ldots + (-1)^{r+s} \omega_1 \wedge \omega_2 \wedge \ldots \wedge \omega_r \wedge \omega_1' \wedge \omega_2' \wedge \ldots \omega_{s-1}' \omega_s'(V)$$

5.1 follows from this formula

q.e.d.

Remark. Another way to prove 5.1 is by induction on r. For 'r=1', 5.1 is a basic property of the inner-product operation. Substitute: $\theta = \theta_1 \wedge \theta_2$, where '$\theta_1$' is of lower gradation than θ, and show that the induction hypothesis works.

Theorem 5.2. Suppose that θ is a differential form of even degree on X and that V be a vector field on X. Set:

$$(\theta)^r = \text{exterior product of r copies of '}\theta\text{'} \tag{5.2}$$

Then:

$$V \lrcorner (\theta)^r = r(V \lrcorner \theta) \wedge (\theta)^{r-1} \tag{5.3}$$

Proof. Suppose that: degree $\theta = 2m$. For $r = 2$:

$$V \lrcorner (\theta)^2 = V \lrcorner (\theta \wedge \theta) = (V \lrcorner \theta) \wedge \theta + (-1)^{\text{degree } \theta} \theta \wedge (V \lrcorner \theta)$$

$$= (V \lrcorner \theta) \wedge \theta + (-1)^{2m} \theta \wedge (V \lrcorner \theta)$$

$$= (V \lrcorner \theta) \wedge \theta + (-1)^{2m(2m-1)} (V \lrcorner \theta) \wedge \theta$$

$$= 2(V \lrcorner \theta) \wedge \theta$$

This proves 5,3 for the case: $r = 2$:. The general case follows readily by induction on r.

q.e.d.

Theorem 5.3. Let the two-form ω define a symplectic structure on the manifold X. The following formula provides the algebraic link between Poisson Bracket determined in terms of vector fields and in terms of exterior derivative and multiplication:

$$n df_1 \wedge df_2 \wedge (\omega)^{n-1} = -\{f_1, f_2\} (\omega)^n, \tag{5.4}$$

where:

$$2n = \dim X. \tag{5.5}$$

and:

$$(\omega)^n = \text{the exterior product of n copies of } \omega \qquad (5.6)$$

Proof. The definition of the Poisson Bracket operation $(f_1, f_2) \rightarrow \{f_1, f_2\}$ given in Section 2 is as follows:

Choose the vector field V_1 such that:

$$df_1 = V_1 \lrcorner \omega \qquad (5.7)$$

Then,

$$\{f_1, f_2\} = V_1(f_2) \qquad (5.8)$$

Since $D^{2n}(X)$ (the set of smooth differential forms of degree 2n) is an $F(X)$-module of dimension one, we have a relation of the following form:

$$df_1 \wedge df_2 \wedge (\omega)^{n-1} = h(\omega)^n, \qquad (5.9)$$

where 'h' is a smooth function. Using 5.8, we have:

$$V_1(f_1) = df_1(V_1) = (V_1 \lrcorner \omega)(V_1)$$

$$= \omega(V_1, V_1) = 0, \qquad (5.10)$$

since ω is a skew-symmetric form,

Apply the operator $(V_1 \lrcorner)$ to both sides of 5.9 and use 5.10:

$$(V_1 \lrcorner) df_1 \wedge df_2 \wedge (\omega)^{n-1} =$$

$$-V_1(f_2) df_1 \wedge (\omega)^{n-1} + df_1 \wedge df_2 \wedge (V_1 \lrcorner (\omega)^{n-1})$$

Bracket for Constraints

$$= -\{f_1, f_2\}df_1 \wedge (\omega)^{n-1} + df_1 \wedge df_2 \wedge (V_1 \, \lrcorner \, (\omega)^{n-1}) \tag{5.11}$$

Now,

$$df_1 \wedge df_2 \wedge (V_1 \, \lrcorner \, (\omega)^{n-1}) = df_1 \wedge df_2 \wedge (V_1 \, \lrcorner \, (\omega \wedge \ldots \wedge \omega))$$

$$= df_1 \wedge df_2 \wedge (V_1 \, \lrcorner \, \omega) \wedge \ldots \wedge \omega + \ldots$$

$$= df_1 \wedge df_2 \wedge df_1 \wedge \ldots \wedge \omega + \ldots$$

$$= 0 + \ldots = 0. \tag{5.12}$$

Combine 5.10 and 5.11:

$$(V_1 \, \lrcorner \,) df_1 \wedge df_2 \wedge (\omega)^{n-1} = -\{f_1, f_2\} df_1 \wedge (\omega)^{n-1} \tag{5.13}$$

Now, use 5.9:

$$V_1 \, \lrcorner \, h(\omega)^n = h\left[(V_1 \, \lrcorner \, \omega) \wedge (\omega)^{n-1} + \omega \wedge (V_1 \, \lrcorner \, \omega) \wedge (\omega)^{n-2} + \ldots\right]$$

$$= h\left[df_1 \wedge (\omega)^{n-1} + \omega \wedge df_1 \wedge (\omega)^{n-2} + \ldots\right]$$

$$= nh \, df_1 \wedge (\omega)^{n-1} \tag{5.13}$$

Comparing 5.11 and 5.12 gives:

$$h = -\{f_1, f_2\},$$

which proves 5.1.

q.e.d.

6. A generalization of the Poisson Bracket for constrained systems using Cartan's formula.

Cartan's formula 5.1 proved in the previous Section enables us to generalize to the constrained case. Let:

$$\theta = \{\theta_1, ..., \theta_m\} \tag{6.1}$$

be an m-tuple of 1-forms on the manifold X of dimension 2n, with a symplectic form ω.

Definition. Let f_1, f_2 be elements of $\mathbf{F}(X)$. Suppose that:

$$2n - m = 2k \tag{6.2}$$

where:

$$k \text{ is a positive integer.} \tag{6.3}$$

Let:

$$n df_1 \wedge df_2 \wedge \theta_1 \wedge ... \wedge \theta_m \wedge (\omega)^{k-1} = -\{f_1, f_2\}_\theta (\omega)^n \tag{6.4}$$

where $\{f_1, f_2\}_\theta$ is the function on X which is defined by 6.4. Then, the operation:

$$(f_1, f_2) \longrightarrow \{f_1, f_2\}_\theta \tag{6.5}$$

is called the **Poisson-Cartan bracket with constraints** defined by the m-tuple of 1-forms 6.1.

In case that 6.2-6.3 is not satisfied, i.e.:

$$2n - m = 2k+1 \tag{6.6}$$

we can modify the construction. Let:

$$\alpha \text{ be a 1-form such that } d\alpha = \omega \tag{6.7}$$

In this case define the Constrained Poisson-Cartan Bracket so that the following identity is satisfied:

$$n df_1 \wedge df_2 \wedge \theta_1 \wedge \ldots \wedge \theta_m \wedge \alpha \wedge (\omega)^{k-1} = - \{f_1, f_2\}_\theta (\omega)^n \tag{6.8}$$

After these generalities, let us turn to study some typical sitautions which occur often in Mechanics.

7. The Poisson-Cartan Bracket on symplectic submanifolds of symplectic manifolds.

Suppose given the following data:

$$\{X, \omega\} \text{ is a symplectic structure on a manifold } X. \tag{7.1}$$

$$Y \text{ is a submanifold of } X. \tag{7.2}$$

$$\omega_Y \text{ is the 2-form on Y which is the restriction of } \omega \text{ to } X. \tag{7.3}$$

In this Section, I will suppose that:

$$\omega_Y \text{ defines a symplectic structure on Y,} \tag{7.4}$$
i.e. ω_Y has no non-zero characteristic vectors.

With 7.4 satisfied, ω_Y defines a Poisson Bracket $\{\,,\,\}_Y$ on smooth functions on Y. However, to compute this Bracket in Local Coodinates for Y requires solving the 'constraint' equations which may define Y. It would then be desirable for computational purposes to have an Algorithm for computing:

$$\{f_Y, f'_Y\}_Y \tag{7.5}$$

for two elements f, f' in **F**(X), where:

$$f_Y = \text{the restriction of f to Y.} \tag{7.6}$$

I will now show that Cartan's Formula provides such an Algorithm. For simplicity, I will only do the following case:

$$\text{codimension Y in X} = 2 \tag{7.7}$$

(The reader should see the possibility of generalizing this case to the general one.)

The Algorithm requires finding two independent 1-forms θ, θ' on X such that:

$$\text{For each } y \in Y, Y_y = \{v \in X_y: \theta(v) = 0 = \theta'(v)\} \tag{7.8}$$

In terms of the notaTion of Section 6, set:

$$\theta = (\theta, \theta') \tag{7.9}$$

Suppose that:

$$\dim(X) = 2n. \tag{7.10}$$

Then, by Definition:

$$ndf \wedge df' \wedge \theta \wedge \theta' \wedge (\omega)^{n-2} = -\{f, f'\}_\theta (\omega)^n \tag{7.11}$$

where '$\{f, f'\}_\theta$' is 'Cartan's Constrained Poisson Bracket'. I will now relate it to the 'unconstrained' Poisson Bracket obtained by restricting ω to Y.

Let us now compare the Bracket Operation: $(f, f') \longrightarrow \{f, f'\}_\theta$ with the bracket 7.5, i.e. the Poisson Bracket relative to {ω restricted to a submanifold} of the restriction of the functions (f_1, f_2) to the submanifold.

Bracket for Constraints

Theorem 7.1. With the above notations, we have:

$$\{f, f'\}_{(\theta, \theta')} \text{ restricted to } Y = \{f_Y, f_Y'\} \tag{7.12}$$

Proof. Let:

$$\omega_Y = \omega \text{ restricted to the submanifold } Y. \tag{7.13}$$

$$V_{f, Y} = \text{the vector field on } Y \text{ such that } df_Y = V_{f, Y} \rfloor \omega_Y \tag{7.14}$$

Let $y \in Y$. We have:

$$V_{f, Y}(y) \rfloor \theta = V_{f, Y}(y) \rfloor \theta' = 0 \tag{7.15}$$

since $V_{f, Y}(y)$ is tangent to Y and 7.8 holds. Now, apply the operator '$V_{f, Y}(y) \rfloor$' to the left-hand side of 7.11.

$$V_{f, Y}(y) \rfloor ndf \wedge df' \wedge \theta \wedge \theta' \wedge (\omega)^{n-2} =$$

$$-nV_{f, Y}(y)(f') \, df \wedge \theta \wedge \theta' \wedge (\omega)^{n-2} + n(n-2)df \wedge df' \wedge \theta \wedge \theta' \wedge [V_{f, Y}(y) \rfloor \omega] \wedge (\omega)^{n-3}$$

$$= -nV_{f, Y}(y)(f') \, df \wedge \theta \wedge \theta' \wedge (\omega)^{n-2} + n(n-2)df \wedge df' \wedge \theta \wedge \theta' \wedge df \wedge (\omega)^{n-3}$$

$$= -nV_{f, Y}(y)(f') \, df \wedge \theta \wedge \theta' \wedge (\omega)^{n-2} + 0 \tag{7.16}$$

Apply the operator '$V_{f, Y}(y) \rfloor$' to the right-hand side of 7.11:

$$V_{f, Y}(y) \rfloor \{f, f'\}_\theta (\omega)^n = \{f, f'\}_\theta n[V_{f, Y}(y) \rfloor \omega] \wedge (\omega)^{n-1}$$

$$= \{f, f'\}_\theta n[df(y)](\omega)^{n-1} \tag{7.17}$$

Equating the right-hand sides of 7.16 and 7.17 proves 7.12.

q.e.d.

CHAPTER 8 LAGRANGE'S EQUATIONS AND EHRESMANN CONNECTIONS IN THE TANGENT BUNDLE

This material in this and the following Chapter was originally intended for Volume 27, and represents another approach to defining a fundamental geometric structure associated with a physical system. In this Chapter. I deal with the unconstrained cases, and treat a simple constrained system in the next Chapter.

1. Introduction.

In Volume 27, I have shown that special cases of Lagrange's Mechanics Equations can be described in the following terms:

Let X be the configuration manifold of the Mechanical System. The Constraints determine the following data

$$\text{A subbundle Z of the tangent vector bundle T(X).} \tag{1.1}$$

$$\text{An Ehresmann Connection for the fiber space Z --> X such that:} \tag{1.2}$$

Then, let us make the following **definition**:

$$\begin{array}{l}\text{A curve } \{t \to x(t)\} \text{ is a \textbf{trajectory} of the}\\ \text{Mechanical System if and only if its tangent vector}\\ \text{field } \{t \to dx/dt(t)\} \text{ lies in the space of horizontal}\\ \text{vectors in T(Z) associated with the Ehresmann Connection}\\ \text{for the fiber space Z --> X}\end{array} \tag{1.3}$$

In Volume 27, I worked via the Levi-Civita connection associated with a Riemannian metric on X. This required that the Mechanical System be of an algebraically special form, albeit one which occurs very frequently in practice. In this Chapter and the following one, I will pursue general situations, starting with the unconstrained systems.

2. Lagrange's unconstrained equations in local coordinates.

Keeping the notations of Section 1, let:

$$\{x^i \varepsilon F(X): 1 \leq i, j, \ldots \leq n\} \tag{2.1}$$

be a coordinate system of smooth functions on X. Set:

$$T(X) = \{(x, v): x \varepsilon X, v \varepsilon X_x\} \tag{2.2}$$
$$\pi: T(X) \longrightarrow X$$
$$\pi(x, v) = x,$$

T(X) is called the **tangent vector bundle to X**.

Define real-valued functions on T(X) via the following formulas:

$$ж^i(x, v) = x^i(x), \text{ i.e.} \tag{2.3}$$
$$\pi^d(x^i) = ж^i.$$

$$v^i(x, v) = dx^i(v) \tag{2.4}$$

One sees readily that the functions:

$$\{ж^1, \ldots, ж^n, v^1, \ldots, v^n\} \tag{2.5}$$

define a coordinate system for T(X). This coordinate system is called the **prolongation** of the given coordinate system 2.1 for X.

A **Lagrangian** is a real-valued function:

$$L: T(X) \longrightarrow R. \tag{2.6}$$

Let such an L be fixed. It defines a function of the real coordinates 2.5. Let:

$$\{L_i, L_{n+i}\} \tag{2.7}$$

be the real-valued functions on T(X) satisfying the following relations:

$$dL = \Sigma_i L_i dx^i + L_{n+i} dv^i, \tag{2.8}$$

Remark. 2.8 impies that the $\{L_i, L_{n+i}\}$ are the partial derivatives of L with respect to the coordinate system 2.5 for T(X).

Definition. The curve $\{t \longrightarrow x(t)\}$ in X is a **trajectory** of the (unconstrained) mechanical system associated with the Lagrangian 2.6 if the following ODE system is satisfied:

$$d\big[L_{n+i}(x(t), dx/dt)\big]/dt = L_i(x(t), dx/dt) \tag{2.9}$$

The second order ODE system 2.9 is called **Lagrange's Equations with Lagrangian L.**

Let us now write 2.9 in a form which presents the second time derivatives of the curve $\{t \longrightarrow x(t)\}$ more explicitly. Introduce the second partial derivatives of the Lagrangian 2.6, denoted as:

$$\{L_{i,j}, L_{i,n+j}, L_{n+i,j}, L_{n+i,n+j}\} \tag{2.10}$$

via the following exterior differential relations:

$$dL_i = \Sigma_j L_{i,j} dx^j + \Sigma_j L_{i,n+j} dv^j, \tag{2.11}$$

$$dL_{n+i} = \Sigma_j L_{n+i,j} dx^i + \Sigma_j L_{n+i,n+j} dv^j, \tag{2.12}$$

We can write the left hand side of Lagrange's ODE 2.9 as follows:

$$dL_{n+i}(\text{evaluated on the tangent vector to the prolonged curve } \{t \longrightarrow (x(t), dx/dt)\} \text{ associated with the curve } \{t \longrightarrow (x(t))\} \tag{2.13}$$

Putting all this together, we have proved:

Theorem 2.1. The curve $\{t \to (x(t))\}$ in X satisfies Lagrange's Equations if and only if it satisfies the following second-order ODE system:

$$\left[\sum_j L_{n+i,n+j}[d^2x^j/dt^2] + \sum_j L_{n+i,j}dx^j/dt - L_i\right](x(t), dx/dt) = 0 \quad (2.14)$$

The coefficient matrices $\{L_{n+i,n+j}\}$, $\{L_{n+i,j}\}$, $\{L_i\}$ in 2.14 have entries which are functions on $T(X)$.

Definition. The Lagrangian L is said to define a **non-singular unconstrained mechanical system on X** if the following condition is satisfied:

> The nxn matrix $\{L_{n+i,n+j}\}$ of functions on $T(X)$ (2.15)
> has an everywhere non-zero determinant.

3. Lagrange's unconstrained equations as the differential equations for straight lines in an Ehresmann connection for the tangent bundle.

Our goal is to interpret the ODE system 2.14 'geometrically' in terms of Ehresmann connections for the Tangent Vector Bundle $\{T(X) \to X\}$. Keep the local coordinate system 2.1 for $T(X)$. The general definition of 'Ehresmann Connection' amounts to the following:

Definition. An Ehresmann connection for the Tangent Vector Bundle $\{T(X) \to X\}$ is defined by a collection:

$$\{\omega^1, ..., \omega^n\} \quad (3.1)$$

of 1-differential forms on $T(X)$ satisfying the following condition:

The collection

$$\{dx^1, ..., dx^n, \omega^1, .., \omega^n\} \quad (3.2)$$

of 1-forms defines a basis for 1-forms, i.e. a Moving Coframe for T(X).

Definition. The Moving Coframe 3.2 for T(X) will be called the **Ehresmann Coframe associated with the Ehresmann connection 3.1** and the local coordinate system $\{x^i\}$ for X. (I will give the standard definition of 'Ehresmann Connection' for general fiber spaces in the next Chapter).

The basic geometric objects defined by an Ehresmann connection in the tangent vector bundle are the 'horizontal curves' and the 'straight lines', which we now define.

Definition. A curve α in T(X) is said to be **horizontal** relative to the connection 3.1 if it is an orbit curve of the Pfaffian system associated with the 1-forms 3.1, i.e. iff:

$$\alpha^d(\omega^1) = 0 = ... = \alpha^d(\omega^n) \quad (3.3)$$

A curve:

$$\{t \dashrightarrow x(t)\} \quad (3.4)$$

in the base space X is said to be a **straight line** relative to the connection 3.1 iff:

The prolonged curve in T(X):

$$\{t \dashrightarrow \alpha(t) = (x(t), dx/dt)(t)\} \quad (3.5)$$

is horizontal relative to the connection 3.1, i.e. conditions 3.3 are satisfied.

Let us now describe more explicitly the differential equations for the straight lines.

4. The differential equations in local coordinates for the straight lines of an Ehresmann connection in the tangent bundle and their equivalence to those of a Lagrangian mechanical system.

Suppose an Ehresmann connection for $\{T(X) \rightarrow X\}$ is given. Let

$$\{x^1, ..., x^n, v^1, ..., v^n\} \tag{4.1}$$

be the prolonged coodinate system 2.1-2.5 for $T(X)$ associated with a coordinate system for X. Let us asume that the connection forms 3.1 are normalized so that they are of the following form:

$$\omega^1 = dv^1 + \Sigma_i \Gamma^1{}_i dx^i$$
$$\cdots \tag{4.2}$$
$$\omega^n = dv^n + \Sigma_i \Gamma^n{}_i dx^i$$

The $n \times n$ matrix $\{\Gamma^j{}_i\}$ of functions on $T(X)$ then defines the connection.

Theorem 4.1. With the connection forms given by 4.2, the straight lines $\{t \rightarrow x(t)\}$ of the connection are defined as the solutions of the following ODE system:

$$d^2 x^i/dt^2 = \Sigma_j \Gamma^i{}_j(x(t), dx/dt)[dx^j/dt] \tag{4.3}$$

where $\{x^i(t)\}$ are the coordinates of the curve $\{t \rightarrow x(t)\}$ in the local coordinate system 2.1 for the manifold X.

Proof. Combine 3.3-3.5 and 4.2.

Let us now introduce a space with variables:

$$\{x^i, v^i, w^i: 1 \leq i, j, \ldots \leq n\} \tag{4.4}$$

Remark. More intrinsically, the space described by the coordinates 4.4 may be considered as the orbit space of the 2-jet space $J^2(R, X)$ under the action of the translation group on the domain space R. This might also be called the **second-order tangent bundle** to X.

We can associate with the system 4.3 the following set of functions of the variables 4.4:

$$\sigma^i = w^i - \Sigma_j \Gamma^i{}_j(x,v) v^j \tag{4.5}$$

The map represented by the functions 4.5 is called the **symbol** of the ODE system 4.3. Similiary, associate with the ODE system 2.4 the map represented by the following set of functions:

$$\tau^i = \Sigma_j L_{n+i,n+j} w^j + \Sigma_j L_{n+i,j} v^j - L_i \tag{4.6}$$

These functions are called the **symbol** of the system 2.14.

Definition. The ODE system 4.3 is said to be **equivalent** to the ODE system 2.4 iff. the following relations hold:

$$\tau^i = \Sigma_j L_{n+i,n+j} \sigma^j \tag{4.7}$$

Theorem 4.1. The ODE Systems 4.3 and 2.14 are equivalent in the above sense iff. the following algebraic relations hold:

$$\Sigma_{jk} L_{n+i,\ n+k} \Gamma^k{}_j v^j = \Sigma_j L_{n+i,\ j} v^j - L_i \tag{4.8}$$

Proof. Substitute 4.5 and 4.6 into 4.7.

Given L, 4.8 may then be regarded as a set of equations for the $\{\Gamma^i_j\}$, which in turn determines an Ehresmann connection whose straight lines are the trajectories of the mechanical system with 'L' as Lagrangian.

Finally for this Chapter (and I plan to continue this analysis further at another point) let us specialize to the most frequently encountered case where:

$$L = 1/2 \sum_{ij} g_{ij}(x) v^i v^j + \sum_i a_i(x) v^i + V(x), \qquad (4.9)$$

where:

$\{g_{ij}\}$ determines the Riemannian metric $\sum_{ij} g_{ij} dx^i dx^j$ on X (4.10)

Remark. Physically, L might represent a 'particle' moving in a 'gravitational' field $\{g_{ij}\}$, and 'electomagnetic' field $\{a_i, V\}$.

We have:

$$L_{n+1,n+j} = g_{ij} \qquad (4.11)$$

$$L_{n+i, j} = \sum_k g_{ik,j}(x) v^k + a_{i,j}(x) v^i + V_{,j}(x) \qquad (4.12)$$

$$L_i = 1/2 \sum_{jk} g_{jk,i}(x) v^j v^k + \sum_j a_{j,i}(x) v^j + V_{,i}(x), \qquad (4.13)$$

(The subcripts ', ()' in 4.12 and 4.13 denote partial derivatives with respect to the $\{x^i\}$ variables.)

Theorem 4.2. With the Lagrangian L of the quadratic form 4.9, 4.8 holds iff.

$$\sum_{jk} g_{ik} \Gamma^k_j v^j = \sum_j \left[\sum_k g_{ik,j}(x) v^k + a_{i,j}(x) v^i + V_{,j}(x) \right] v^j$$
$$- 1/2 \sum_{jk} g_{jk,i}(x) v^j v^k + \sum_j a_{j,i}(x) v^j + V_{,i}(x), \qquad (4.14)$$

Thus, 4.14 consititutes a set of algebraic equations to be solved for the functions:

$$\left\{ \sum_j \Gamma^i_j(x, v) v^j \right) \qquad (4.15)$$

Proof. Substitute 4.11-4.13 into 4.8.

q.e.d.

Remark. I hope to study the equations 4.14 - and similiar systems - in more detail at a later point.

CHAPTER 9 - EHRESMANN CONNECTIONS IN DIFFERENTIABLE FIBER SPACES AND CONSTRAINED MECHANICAL SYSTEMS

1. Introduction.

In Volume 27, I have shown that the notion of 'affine connection' as a covariant differentiation operation is a very convenient formalism for the description of the analytical properties of Path Systems which occur naturally in mechanics problems. I will now show how the Ehresmann notion of 'connection in fiber spaces' is also well-adapted to study the geometry of such systems.

Here is one basic geometric insight which the Ehresmann theory gives us. In unconstrained mechanics problems, the trajectory curves:

$$t \longrightarrow x(t) \; \varepsilon \; X \tag{1.1}$$

of many mechanical systems possess the following goemetric structure:

> There is a connection in the Ehresmann sense for (1.2)
> the tangent vector bundle $\pi: T(X) \longrightarrow X$ such that the
> trajectories in X are the projection of the curves in
> T(X) which are: **a)** horizontal relative to the Ehresmann
> connection and **b)** also are tangent vector curves to curves
> in X.

Constrained Mechanics problems generalize this structure in the following way:

There is a submanifold:

$$Z \subset T(X) \qquad (1.3)$$

such that:

$$\pi \text{ restricted to } Z \text{ defines a fiber space over } X. \qquad (1.4)$$

There is an Ehresmann connection for the fiber space: (1.5)

$$\pi: Z \longrightarrow X$$

The trajectories of the mechanical system are the projection of the curves: $\{t \longrightarrow z(t)\}$ in Z which satisfy the following conditions:

$$\{t \longrightarrow z(t)\} \text{ is the tangent vector curve to} \qquad (1.6)$$
the projected curve $\{t \longrightarrow \pi z(t)\}$

$$\{t \longrightarrow z(t)\} \text{ is horizontal relative to the} \qquad (1.7)$$
Ehresmann connection for the fiber
space $Z \longrightarrow X$.

From Cartan's point of view, there is an even more general way of putting this:

One is given a manifold Z and two Pfaffian systems (1.8) on Z. The trajectories are the curves which are orbit curves of both Pfaffian systems.

This geometric point of view also lends itself to generalization to field-theoretic and continuum problems, a topic I hope to get to at a later point.

2. Differentiable fiber spaces and the foliation by vertical vectors.

The 'Differentiable Fiber Spaces' are basic differential-topological objects in Ehresmann's framework. I will now recall some of their basic definitions and properties.

Definition. Let:

$$\pi: Z \dashrightarrow X \qquad (2.1)$$

be a smooth map between smooth manifolds which is a submersion, i.e. is such that:

$$\pi_*(T(Z)) = T(X) \qquad (2.2)$$

where

$$\pi_*: T(Z) \dashrightarrow T(X)$$

is the linear map of tangent bundles induced by the given map π. The triple (π, Z, X) will be called a **differentiable fiber space**.

For $z \in Z$, let:

$$V_\pi(z) = \{v \in Z_z: \pi_*(v) = 0\} \qquad (2.3)$$

Then, because 2.2 is satisfied, we have:

$$\dim(V_\pi(z)) = \dim Z - \dim X. \qquad (2.4)$$

The vectors in $V_\pi(z)$ are said to be **vertical** with respect to π. Because of 2.4, we have:

> The assignment $z \dashrightarrow V_\pi(z)$ defines a distribution on Z, called the **vertical distribution relative to** π. $\qquad (2.5)$

A tangent vector field V: z --> V(z) on Z is said to be **vertical relative to** π if:

$$V(z) \, \varepsilon \, \mathbf{V}_\pi(z) \text{ for all } z \, \varepsilon \, Z. \tag{2.6}$$

Condition 2.6 is equivalent to the following:

$$\pi_*(V) = 0, \text{ i.e. } V(\pi^*(f)) = 0 \text{ for all } f \, \varepsilon \, \mathbf{F}(X) \tag{2.7}$$

Set:

$$\mathbf{V}_\pi = \{V \varepsilon \, \mathbf{U}(Z) \colon \pi_*(V) = 0\} \tag{2.8}$$

The elements of $\mathbf{V}_\pi$ are called the **vertical vector fields**. Then:

$$[\mathbf{V}_\pi, \mathbf{V}_\pi] \subset \mathbf{V}_\pi \tag{2.9}$$

i.e. $\mathbf{V}_\pi$ is a Frobenius Integrable vector field system on Z.

Thus, we see that a differentiable fiber space determines a Frobenius Integrable vector field system (and hence a foliation), thought of as 'vertical'. Ehresmann's basic geometric idea was to define a 'connection' as a complementary distribution, thought of as 'horizontal'.

3. Ehresmann connections and horizontal liftngs of curves and vector fields in the base.

Let:

$$(\pi, Z, X) \tag{3.1}$$

be a differentiable fiber space.

Ehresmann Connections

Definitions. An **Ehresmann Connection** for the fiber space (π, Z, X) is defined by a smooth distribution :

$$z \longrightarrow H(z) \subset Z_z \qquad (3.2)$$

of linear subspaces of the tangent bundle to Z such that:

$$Z_z = V_\pi(z) \oplus H(z) \text{ for all } z \, \varepsilon \, Z. \qquad (3.3)$$

3.3 implies the following condition:

$$U(Z) = V_\pi \oplus H, \qquad (3.4)$$

where **H** is the vector field system associated with the distribution $z \longrightarrow H(z)$, i.e.:

$$H = \{V \, \varepsilon \, U(Z) : V(z) \, \varepsilon \, H(z)\} \text{ for all } z \, \varepsilon \, Z\}$$

Let us now recall some Lie-theoretic terminology.

Definition. Let $\pi: Z \longrightarrow X$ be a smooth map between manifolds. A vector field V on Z is said to be **projectable relative to** π if there is a vector field W on X such that the following condition is satisfied:

$$\pi_*(V(z)) = W(\pi(z)) \text{ for all } z \, \varepsilon \, Z \qquad (3.5)$$

Condition 3.5 is equivalent to the following:

$$\pi^*(W(f)) = V(\pi^*(f)) \text{ for all } f \, \varepsilon \, F(X) \qquad (3.6)$$

Ehresmann Connections

Returning to the Ehresmann theory of connections in fiber spaces, let H be an Ehresmann connection for the fiber space $\pi: Z \to X$. Let:

$H(\pi)$ = set of vector fields in H which are projectable under π, i.e. suppose that for each $V \in H(\pi)$ the following condition is satisfied: (3.7)

There is a vector field V' in X such that $\pi_*(V) = V'$, i.e. (3.7a)
$\pi_*(V(z)) = V'(\pi(z))$ for all $z \in Z$.

Theorem 3.1. There is a unique map:

$$L(\pi, H) : U(X) \to H(\pi) \quad (3.8)$$

such that:

$L(\pi, H)$ is R-linear. (3.9)

For each vector field V in $U(X)$, the element $L(\pi, H)(V)$ of $U(Z)$ is projectable under the map π. (3.9a)

$\pi_* L(\pi, H)$ = identity (3.10)

$L(\pi, H)(fV) = \pi^*(f) L_\pi(V)$
for $f \in F(X), V \in U(X)$ (3.11)

$L(\pi, H)$ is called the **horizontal lifting map on vector fields** associated with the connection H.

Proof. We see from 2.4 and 2.3 that the map:

$$\pi_* \text{ (restricted to } H) : H(\pi) \to U(X) \quad (3.12)$$

is one-one. To prove it is onto notice that the maps $\pi_*: Z_z \rightarrow X_{\pi(z)}$ are vector space isomorphisms, for all $z \in Z$. Thus, given $W \in U(X)$, we can define a $V \in H$ such that $\pi_*(V) = W$ by the rule:

$$V(z) = \pi_*^{-1}(W(\pi(z))). \tag{3.13}$$

$L(\pi, H)$ can now be chosen as the inverse of π_* restricted to H, i.e.

$$L(\pi, H)(W) = V. \tag{3.14}$$

q.e.d.

4. Curvature for Ehresmann connections.

Having defined an Ehresmann connection as a horizontal distribution, it is geometrically natural to think that the first-order geometric 'invariants' of this distribution are basic objects. In this case, these invariants would measure the Frobenius Integrability of the distribution; they are called the 'curvature'.

Continue with the notations of Section 3. H is the horizontal distribution defining an Ehresmann connection.

Theorem 4.1. Let:

$$K: U(X) \times U(X) \longrightarrow V_\pi \tag{4.1}$$

be the map defined as follows:

$$K(V, V') = [L(\pi, K)(V), L(\pi, K)(V')] \text{ projected} \tag{4.2}$$
$$\text{into } V_\pi \text{ with respect to the direct-sum}$$
$$\text{decomposition 3.4.}$$

Then, K satisfies the following conditions:

$$K \text{ is R-bilinear and skew-symmetric.} \tag{4.3}$$

$$\mathbf{K}(f\mathbf{V}, \mathbf{V'}) = \pi^*(f)\mathbf{K}(\mathbf{V}, \mathbf{V'}) = \mathbf{K}(\mathbf{V}, f\mathbf{V'}) \qquad (4.4)$$
$$\text{for } f \in \mathbf{F}(X); \mathbf{V}, \mathbf{V'} \in \mathbf{U}(X)$$

Proof. 4.3 follows from the fact that the Jacobi-Lie bracket operation [,] on vector fields is R-bilinear and skew-symmetric. The projection from $\mathbf{U}(Z)$ to V_π is $\mathbf{F}(Z)$-linear.

To prove 4.4, notice that, although the Jacobi-Lie bracket operation [,] is not $\mathbf{F}(Z)$-linear, when calculating [fV, V'] the term which violates the $\mathbf{F}(Z)$-linearity is V'(f)V. This becomes zero when projected into $\mathbf{U}_\pi$, for V, V' $\in$ H, thus proving 4.4.

q.e.d.

Definition. K defined by formula 4.4 is called the **curvature tensor of the Ehresmann connection defined by the vector field system H.**

Remark. The vanishing of the curvature tensor **K** defined by 4.4 is then the condition that the horizontal connection distribution **H** be Frobenius Integrable.

5. The horizontal lifting of curves and vector fields relative to an Ehresmann connection and parallel transport.

A key geometric property of the Ehresmann connections is that they define a 'horizontal lifting of curves' and associated 'parallel-transport' in a very elegant and general geometric way. Let us briefly recall the definition of these two operations.

Definition. Let:

$$\{t \to x(t), a \leq t \leq b\} \tag{5.1}$$

be a continuous, piecewise-smooth curve in the base space X of the fiber space (π, Z, X) with a connection **H**. A curve $t \to z(t)$ in Z is said to be a **horizontal lifting** of the curve 5.1 iff:

$$\pi(z(t)) = x(t) \text{ for } a \leq t \leq b \tag{5.2}$$

$$dz/dt \; \varepsilon \; \mathbf{H}(z(t)) \tag{5.3}$$

Theorem 5.1. With the above notations, given a point:

$$z_0 \; \varepsilon \; \pi^{-1}(x(a)) \tag{5.3}$$

in the fiber of the map π above the initial point of the base- curve 5.1, there is at most one horizontal lifting:

$$\{t \to z(t)\} \tag{5.5}$$

of the curve 5.1 such that:

$$z(a) = z_0. \tag{5.6}$$

Locally about z_0, the horizontally-lifted curve 5.5 is the solution of a system of ordinary differential equations of 'control' type with the coordinates of the base curve 5.2 as 'inputs'.

Proof. To see this we must introduce local coordinate systems for Z and X about the points z_0 and $z(a)$ which are 'adapted' to the fiber space map π. Using the Implicit Function Theorem and the hypothesis that the Jacobian of the map π is of maximal rank at every point of Z, we deduce that there are local coordinate functions labelled:

$$(z^1, ..., z^n), n = \dim Z \tag{5.7}$$

$$(x^1, ..., x^m), \; m = \dim X \tag{5.8}$$

defined in a neighborhood of z_0 and $x(a)$ respectively such that:

$$\pi^*(x^1) = z^1, ..., \pi^*(x^m) = z^m \tag{5.9}$$

Our hypotheses about **H** imply that there are one-differential forms:

$$\theta^{m+1} = dz^{m+1} - \Gamma^{m+1}{}_1 dz^1 - \Gamma^{m+1}{}_2 dz^2 - ... - \Gamma^{m+1}{}_m(z) dz^m$$
$$...$$
$$\theta^n = dz^n - \Gamma^n{}_1 dz^1 - \Gamma^n{}_2 dz^2 - ... - \Gamma^n{}_m(z) dz^m \tag{5.10}$$

defined in the neighborhood of z_0 such that 5.11 and 5.12 are satisfied:

> The matrix $\{\Gamma^i{}_j : m+1 \le i \le n, \; 1 \le j \le m\}$ which appears in 5.10 consists of smooth functions defined in a neighborhood of x_0. The coefficient functions $f^i{}_j$ can be described as functions $\Gamma^i{}_j(z^1, ..., z^n)$ of the indicated variables. \hfill (5.11)

> A tangent vector v to Z in this neighborhood lies in the horizontal distribution **H** iff:

$$\theta^{m+1}(v) = 0 = ... = \theta^n(v) \tag{5.12}$$

In more algebraic terms 5.12 means that:

> The Pfaffian system:

$$\theta^{m+1} = 0 = ... = \theta^n \tag{5.13}$$

> is dual to the horizontal vector field system **H**.

Suppose that the curve 5.1 in the base space X has coordinate functions:

$$\{x^1(t), ..., x^m(t)\} \tag{5.14}$$

The conditions that a curve $\{t \to z(t)\}$ in Z be a horizontal lifting of 5.1 are then that the following system of ordinary differential equations be satisfied:

$$\begin{aligned}dz^{m+1}/dt = &\ \Gamma^{m+1}{}_1[x^1(t), ..., x^m(t), z^{m+1}(t), ..., z^n(t)]dx^1/dt \\ &+ \Gamma^{m+1}{}_2[x^1(t), ..., x^m(t), z^{m+1}(t), ..., z^n(t)]dx^2/dt \\ &+ ... \\ &+ \Gamma^{m+1}{}_m[x^1(t), ..., x^m(t), z^{m+1}(t), ..., z^n(t))dx^m/dt\end{aligned} \tag{5.15}$$

...

$$\begin{aligned}dz^n/dt = &\ \Gamma^n{}_1[x^1(t), ..., x^m(t), z^{m+1}(t), ..., z^n(t)]dx^1/dt \\ &+ \Gamma^n{}_2[x^1(t), ..., x^m(t), z^{m+1}(t), ..., z^n(t)]dx^2/dt \\ &+ ... \\ &+ \Gamma^n{}_m[x^1(t), ..., x^m(t), z^{m+1}(t), ..., z^n(t))dx^m/dt\end{aligned}$$

Theorem 5.1 now follows from standard existence and uniqueness theorems about systems of smooth ODE's of the form 5.15, with the $x^1(t), ..., x^m(t)$ given as smooth functions of t. As for the 'control' aspects, notice that the base-curve $\{t \to (x^1(t), ..., x^m(t)\}$ plays the role of the 'control'.

q.e.d.

Let us now define the operation of parallel transport. Let:

$$\{t \to x(t): a \leq t \leq b\} \tag{5.16}$$

again be a piece-wise smooth, continuous curve in the base X. Then, given a point:

$$z_0 \, \varepsilon \, \pi^{-1}(x(a)), \tag{5.17}$$

there is at most one unique horizontal lifting:

$$\{t \rightarrow z(t): a \leq t \leq b\} \tag{5.18}$$

of the curve 5.15 such that:

$$z(a) = z_0.$$

(The proof of this fact involves the uniqueness part of the standard Picard-type basic ODE Existence-Uniqueness Theorem.)

Definition. For each t such that $a \leq t \leq b$, the point $z(t)$ of $\pi^1(x(t))$ is the **parallel transport relative to the Ehresmann connection H** of the point z_0 along the base-curve 5.16.

6. Horizontal completeness.

The system 5.15 of ODE's that determines the horizontal lifting is - in general - nonlinear, hence may not admit solutions for the whole parameter interval $\{a \leq t \leq b\}$. We would like to focus attention on the class of connections which do not have this pathology. The following definition gives this class a name:

Definition. The Ehresmann connection H is said to be **horizontally complete** if each curve $\{t \rightarrow x(t), a \leq t \leq b\}$ in X and each $z_0 \; \varepsilon \; \pi^{-1}(q(a))$ admits a horizontal lifting $\{t \rightarrow z(t), a \leq t \leq b\}$ such that: $x(a) = x_0$.

It will be convenient to express 'horizontal completness' in terms of properties of vector fields.

Definition A vector field V on a manifold Z is said to be **complete** if, for each point $z_0 \; \varepsilon \; Z$, there is a smooth curve:

$$\{t \rightarrow z(t): t \; \varepsilon \; R\} \tag{6.1}$$

such that:

$$z(0) = z_0 \tag{6.2}$$

and:

$$dz/dt = V(z(t)) \tag{6.3}$$

The curve satisfying 6.1-6.3 is called the **orbit curve** of V beginning at the point x_0.

Geometrically, the condition of 'completeness' means that each 'small' piece of an orbit curve of V can be extended to an orbit curve defined over the entire real line. We also use the following notation:

$$z(t) = \exp(tV)(z_0) \tag{6.4}$$

The mapping:

$$t \longrightarrow \exp(tV) \tag{6.5}$$

is then a homomorphism of the additive Lie group of real numbers into the group of diffeomorphisms of Z, called the **one-parameter diffeomorphism group on Z generated by the vector field V**.

Definition. The connection **H** for the fiber space is said to be **horizontally complete** if the following condition is satisfied:

$$\text{For each complete vector field V in the base space X, the horizontally lifted curve } \mathbf{L}(\pi, \mathbf{H})(V) \text{ on Z is also complete.} \tag{6.6}$$

The following result describes a main geometric property of the horizontally-lifted vector fields.

Theorem 6.1. Suppose that the connection **H** is horizontally complete in the above sense. Then, for each complete vector field V on the base space X, the one-parameter group:

$$t \longrightarrow \exp(t\mathbf{L}(\pi, \mathbf{H})(V)) \tag{6.7}$$

of diffeomorphisms of Z generated by its horizontal lifting has the following property:

For each $t \in R$, each $x \in X$,

$$\exp(t\mathbf{L}(\pi, \mathbf{H})(V))(\pi^{-1}(x)) = \pi^{-1}(\exp(tV)x) \tag{6.8}$$

In words, 6.8 expresses the property that:

The one-parameter group 6.8 maps fibers of π into fibers.

Proof. Follows readily from the definitions, hence is left to the reader.

Another way of expressing the geometric property of 'horizontal lifting' is in terms of 'projectability' of vector fields, a notion defined in Section 3:

$L(\pi, \mathbf{H})(V))$ is projectable under π to the vector field V in the base X. (6.9)

7. Path systems defined by Ehresmann connections on sub-bundles of the tangent bundle.

Let X continue as a smooth manifold; for example, the configuration space of a mechanical system. Let:

$$T(X) = \{(x, v): x \in X, v \in X_x\} \quad (7.1)$$

be the tangent vector bundle to X. The fiber space map:

$$\pi: T(X) \longrightarrow X \quad (7.2)$$

is the Cartesian projection map:

$$(x, v) \longrightarrow x \quad (7.3)$$

Let:
$$\mathbf{W} \subset \mathbf{V}(X) \quad (7.4)$$

be a non-singular, smooth vector field system on X, which will define the constraints of a mechanical systems. For $x \in X$, let:

$$\mathbf{W}(x) = \{V(x): V \in \mathbf{W}(X)\} \quad (7.5)$$

be the set of tangent vectors to x which are the restriction to x of the vector fields in $\mathbf{W}$. Then, the assignment:

$$x \longrightarrow \mathcal{W}(x) \tag{7.6}$$

defines a distribution on X in the sense of Chevalley. Also, set:

$$T(X)_{\mathcal{W}} = \{(x, v): x \in X, v \in \mathcal{W}(x)\} \subset T(X) \tag{7.7}$$

$T(X)_{\mathcal{W}}$ defines a **sub-bundle** of the tangent vector bundle to X.

If $t \longrightarrow x(t)$ is a piece-wise smooth curve in X, let:

$$t \longrightarrow dx(t)/dt \, \varepsilon \, X_{x(t)} \tag{7.8}$$

be its tangent vector field. We can define the following curve in T(X):

$$t \longrightarrow (x(t), dx(t)/dt) = v(t) \tag{7.9}$$

The curve 7.9 in T(X) is said to be the **tangent prolongation** to the curve in X. Then, from these definitions, we have:

> The curve $t \longrightarrow x(t)$ in X is an orbit curve of the vector field system $\mathcal{W}$ if and only if its tangent prolongation $t \longrightarrow (x(t), dx(t)/dt)$ satisfies the following condition:
>
> $$t \longrightarrow (x(t), dx(t)/dt) \text{ lies in } T(X)_{\mathcal{W}} \tag{7.10}$$

Now, if 7.10 is satisfied, we can impose further conditions on the curve. A geometrically natural way of doing this is to suppose given another vector field system on $T(X)_{\mathcal{W}}$ and require that the prolonged curve $t \longrightarrow (x(t), dx(t)/dt)$ be an orbit curve of this vector field system. An Ehresmann connection for the fiber space:

$$\begin{aligned} T(X)_{\mathcal{W}} &\longrightarrow X \\ (x, v) &\longrightarrow x \end{aligned} \tag{7.11}$$

gives such a system.

Definition. Let :

$$H \subset \mathbb{U}(T(X)_\mathbb{W}) \qquad (7.12)$$

be a vector field system on the sub-bundle of the tangent vector bundle to the configuration manifold X associated with the constraint system **W**. Suppose that:

> H defines an Ehresmann connection for the fiber (7.13)
> space map: $T(X)_\mathbb{W} \to X$.

Then, a smooth curve $t \to x(t)$ in X is a **straight line for the connection H** if the following condition is satisfied:

> The prolonged curve $\{t \to (x(t), dx/dt) = v(t))\}$ (7.14)
> lies in $T(X)_\mathbb{W}$ and is horizontal relative to
> the vector field system **H**.

Theorem 7.1. The differential equations for the curves $t \to x(t)$ in X that are straight lines of **H** are second order, of the following form:

$$\theta(dv/dt) = 0 \qquad (7.15)$$
for each 1-form θ on $T(X)_\mathbb{W}$ which satisfies the
condition: $\theta(H) = 0$.

Proof. 7.15 is the expression of 7.14 in dual terms. As for 'second order', note that the prolonged curve $\{t \to (x(t), dx/dt) = v(t))\}$ involves the firts derivatives of the curve $\{t \to x(t)\}$, while 7.15 involves a first-order condition on $\{t \to (x(t), dx/dt) = v(t))\}$, which therefore impies a second-order condition on the curve $\{t \to x(t)\}$.

q.e.d.

I will now 'switch gears' and work in Local Coodinates, dealing with one of the simplest Constraint situations.

8. The constrained Lagrange Equations in Local Coordinates, in the case that the constraint subbundle is of codimension one.

Again, I will suppose that:

X is a manifold of dimension n,

and make the following simplifying assumptions about the constraint equations:

$\{\pi_Z: Z \to X\}$ is a sub-bundle of the Tangent Vector Bundle (8.1)

$\{\pi: T(X) \to X\}$ such that:

The fibers $\pi_Z^{-1}(x)$ of the map π_Z are linear subspaces of (8.2) codimension one of the tangent space X_x.

To define 'constrained mechanics' in this situation, let us use local coordinates:

$$\{x^i, v^i: 1 \leq i, j, k, \ldots \leq n\} \tag{8.3}$$

for T(X) such that:

The submanifolds $\{x^i = \text{constant}\}$ are the fibers of the (8.4) fiber bundle projection map: $T(X) \to X$, i.e. the $\{x^i\}$ are the 'base-like' coordinates, or the 'lifting' to $T(X)$ of local coordinates on X. The $\{v^i\}$ are the differentials of the local coordinates on X, considered as linear functions on the tangent spaces. Physically, if 'X' is the 'configuration space', then $\{x^i\}$ are the 'psition coordinates' and $\{v^i\}$ are the 'velocity coordinates'.

Suppose that:

$$\phi(x, v) = \Sigma_i \phi_i(x) v^i \tag{8.5}$$

is a smooth real-valued function on T(X) which is linear on the fibers of the map {T(X) --> X}, i.e.:

ϕ is a 1-differential form on X, considered as (8.6)
a linear, real-valued function on tangent vectors.

Set:

$$Z = \{(x, v) \in T(X): \phi(x, v) = 0\}. \qquad (8.7)$$

Let us suppose given the following data:

A Lagrangian function L: T(X) --> R, (8.8)
{(x, v) --> L(x, v).

As in the previous Chapter, denote the partial derivatives of L with respect to the coordinates 8.3, by means of subscripts, i.e.:

$$dL = \Sigma_i L_i dx^i + \Sigma_i L_{n+i} dv^i \qquad (8.9)$$

$$dL_j = \Sigma_i L_{ji} dx^i + \Sigma_i L_{j,\,n+i} dv^i \qquad (8.10)$$

. . .

An Ehresmann Connection for the Tangent Bundle {T(X) --> X}, defined by 1-forms $\{\omega^i\}$ on T(X) such that:

$$\omega^i = dv^i + \Sigma_j \Gamma^i_j dx^j \qquad (8.11)$$

In the previous Chapter, we have dealt with relations betwee the unconstrained trajectory curves of the Lagrangian L and straight lines of an Ehresmann connection for the tangent bundle to X. Now let us turn to study the trajectories of the Mechanical System $\{L, \phi\}$ with Lagrangian L and constraints 8.7:

Definition. A curve $\{t \to x(t)\}$ in the base space X is said to be a **trajectory** of the constrained mechanical system $\{L, \phi\}$ if there is a real-valued function:

$$\{t \to \lambda(t)\} \tag{8.12}$$

such that 8.13 and 8.14 are satisfied:

$$\left(d[L_{n+i}]/dt - L_i\right)(x(t), dx/dt) = \lambda(t)\phi_i(x(t)) \tag{8.13}$$

$$\Sigma_i \phi_i(x(t))[dx^i/dt] = 0 \tag{8.14}$$

Remark. See my book "Geometry, Physics and Systems" for a more complete description of the background of these Equations of Constrained Motion.

Theorem 8.1. A curve $\{t \to (x(t), \lambda(t))\}$ is a trajectory of the Constrained Mechanical System $\{L, \phi\}$ if and only if the following second order ODE system is satisfied:

$$\left(\Sigma_j L_{n+i,\,n+j}[d^2x^j/dt^2] + \Sigma_j L_{n+i,j}[dx^j/dt] - L_i\right) - \lambda(t)\phi_i(x(t)) = 0 \tag{8.15}$$

$$\Sigma_i \phi_i(x(t))[dx^i/dt] = 0 \tag{8.16}$$

Proof. As was done in the previous Chapter, we can work out the left-hand side of 8.13 in terms of the first and second derivatives of the curve $\{t \to x(t)\}$. The left hand side of 8.15 is the result.

q.e.d.

Let us again introduce the variables:

$$\{x^i, v^i, w^i, \lambda\} \tag{8.17}$$

and in this space the hypersurface Z defined by the relation:

$$\sum_i \phi_i(x) v^i = 0 \qquad (8.18)$$

Consider the following functions of the variables 8.17, which determine the 'symbol' of the ODE system 8.15-8.16:

$$\sum_j L_{n+i,\, n+j} w^j + \sum_j L_{n+i,\, j} v^j - L_i - \lambda(t)\phi_i(x(t)) \qquad (8.15)$$

$$\sum_i \phi_i v^i \qquad (8.16)$$

Given the connection 8.11, write down the following functions:

$$\sum_j L_{n+i,\, n+j} w^j + \sum_{jk} L_{n+i,\, n+j} \Gamma^j_k v^k \qquad (8.17)$$

Definition. The Constrained Mechanical System $\{L, \phi\}$ is said to be **trajectory-equivalent** to the Ehresmann connection 8.11 iff. the following realtions are satisfied:

$$\sum_j L_{n+i,\, j} v^j - L_i = \sum_{jk} L_{n+i,\, n+j} \Gamma^j_k v^k \qquad (8.18)$$

for all points of the space 8.17 such that:

$$\sum_i \phi_i v^i = 0. \qquad (8.19)$$

Here is the main result of this Chapter:

Theorem 8.1. If condition 8.18-8.19 is satisfied, then each trajectory of the Constrained Mechanical System $\{L, \phi\}$ is a straight line of the Ehresmann Connection 8.11.

Proof. If 8.18-8.19 is satisfied we see from the above work that, given a curve $\{t \to x(t)\}$ in X, it satisfies the ODE system iff. it satisfies the ODE system.

q.e.d.

What must now be done is to work out some conditions on the connection 8.11 implied by 8.18-8.9; generalize to more constraints; work out the conditions that the curvature of the connection vanish, etc. I will leave this considerable 'Programm' to another book!